GLOBAL
WARMING

Fossil Fuels
and Pollution

The Future of Air Quality

Julie Kerr Casper, Ph.D.

Facts On File
An imprint of Infobase Publishing

FOSSIL FUELS AND POLLUTION: The Future of Air Quality

Facts On File, Inc.
An imprint of Infobase Publishing
132 West 31st Street
New York NY 10001

Library of Congress Cataloging-in-Publication Data
Casper, Julie Kerr.
    Fossil fuels and pollution : the future of air quality / Julie Kerr Casper.
        p. cm.—(Global warming)
    Includes bibliographical references and index.
    ISBN 978-0-8160-7265-1
    1. Fossil fuels—Environmental aspects—Juvenile literature. 2. Pollution—Environmental aspects—Juvenile literature. 3. Environmental protection—Juvenile literature. I. Title
    TP 318.3.C37 2010
    363.738′74—dc22   2009012612

Facts On File books are available at special discounts when purchased in bulk quantities for businesses, associations, institutions, or sales promotions. Please call our Special Sales Department in New York at (212) 967-8800 or (800) 322-8755.

You can find Facts On File on the World Wide Web at http://www.factsonfile.com

Text design by Erik Lindstrom
Illustrations by Dale Williams
Composition by Hermitage Publishing Services
Cover printed by Bang Printing, Brainerd, MN
Book printed and bound by Bang Printing, Brainerd, MN
Date printed: January 2010
Printed in the United States of America

10 9 8 7 6 5 4 3 2 1

This book is printed on acid-free paper.

CONTENTS

PREFACE

We do not inherit the Earth from our ancestors—
we borrow it from our children.

This ancient Native American proverb and what it implies resonates today as it has become increasingly obvious that people's actions and interactions with the environment affect not only living conditions now, but also those of many generations to follow. Humans must address the effect they have on the Earth's climate and how their choices today will have an impact on future generations.

Many years ago, Mark Twain joked that "Everyone talks about the weather, but no one does anything about it." That is not true anymore. Humans are changing the world's climate and with it the local, regional, and global weather. Scientists tell us that "climate is what we expect, and weather is what we get." Climate change occurs when that average weather shifts over the long term in a specific location, a region, or the entire planet.

Global warming and climate change are urgent topics. They are discussed on the news, in conversations, and are even the subjects of horror movies. How much is fact? What does global warming mean to individuals? What should it mean?

The readers of this multivolume set—most of whom are today's middle and high school students—will be tomorrow's leaders and scientists. Global warming and its threats are real. As scientists unlock the mysteries of the past and analyze today's activities, they warn that future generations may be in jeopardy. There is now overwhelming evidence that human activities are changing the world's climate. For thousands of years, the Earth's atmosphere has changed very little; but today, there are problems in keeping the balance. Greenhouse gases are being added to the atmosphere at an alarming rate. Since the Industrial Revolution (late 18th, early 19th centuries), human activities from transportation, agriculture, fossil fuels, waste disposal and treatment, deforestation, power stations, land use, biomass burning, and industrial processes, among other things, have added to the concentrations of greenhouse gases.

These activities are changing the atmosphere more rapidly than humans have ever experienced before. Some people think that warming the Earth's atmosphere by a few degrees is harmless and could have no effect on them; but global warming is more than just a warming—or cooling—trend. Global warming could have far-reaching and unpredictable environmental, social, and economic consequences. The following demonstrates what a few degrees' change in the temperature can do.

The Earth experienced an ice age 13,000 years ago. Global temperatures then warmed up 8.3°F (5°C) and melted the vast ice sheets that covered much of the North American continent. Scientists today predict that average temperatures could rise 11.7°F (7°C) during this century alone. What will happen to the remaining glaciers and ice caps?

If the temperatures rise as leading scientists have predicted, less freshwater will be available—and already one-third of the world's population (about 2 billion people) suffer from a shortage of water. Lack of water will keep farmers from growing food. It will also permanently destroy sensitive fish and wildlife habitat. As the ocean levels rise, coastal lands and islands will be flooded and destroyed. Heat waves could kill tens of thousands of people. With warmer temperatures, outbreaks of diseases will spread and intensify. Plant pollen mold spores in the air will increase, affecting those with allergies. An increase in severe weather could result in hurricanes similar or even stronger than Katrina in 2005, which destroyed large areas of the southeastern United States.

Higher temperatures will cause other areas to dry out and become tinder for larger and more devastating wildfires that threaten forests, wildlife, and homes. If drought destroys the rain forests, the Earth's delicate oxygen and carbon balances will be harmed, affecting the water, air, vegetation, and all life.

Although the United States has been one of the largest contributors to global warming, it ranks far below countries and regions—such as Canada, Australia, and western Europe—in taking steps to fix the damage that has been done. Global Warming is a multivolume set that explores the concept that each person is a member of a global family who shares responsibility for fixing this problem. In fact, the only way to fix it is to work together toward a common goal. This seven-volume set covers all of the important climatic issues that need to be addressed in order to understand the problem, allowing the reader to build a solid foundation of knowledge and to use the information to help solve the critical issues in effective ways. The set includes the following volumes:

Climate Systems
Global Warming Trends
Global Warming Cycles
Changing Ecosystems
Greenhouse Gases
Fossil Fuels and Pollution
Climate Management

These volumes explore a multitude of topics—how climates change, learning from past ice ages, natural factors that trigger global warming on Earth, whether the Earth can expect another ice age in the future, how the Earth's climate is changing now, emergency preparedness in severe weather, projections for the future, and why climate affects everything people do from growing food, to heating homes, to using the Earth's natural resources, to new scientific discoveries. They look at the impact that rising sea levels will have on islands and other areas worldwide, how individual ecosystems will be affected, what humans will lose if rain forests are destroyed, how industrialization and pollution puts peoples' lives at risk, and the benefits of developing environmentally friendly energy resources.

The set also examines the exciting technology of computer modeling and how it has unlocked mysteries about past climate change and global warming and how it can predict the local, regional, and global climates of the future—the very things leaders of tomorrow need to know *today.*

> *We will know only what we are taught;*
> *We will be taught only what others deem is important to know;*
> *And we will learn to value that which is important.*
> —Native American proverb

ACKNOWLEDGMENTS

Global warming may very well be one of the most important issues facing you in your lifetime. The decisions you make on energy sources and daily conservation practices will determine not only the quality of your life, but also those of your descendants.

I cannot stress enough how important it is to gain a good understanding of global warming: what it is, why it is happening, how it can be slowed down, why everybody is contributing to the problem, and why *everybody* needs to be an active part of the solution.

I would sincerely like to thank several of the federal agencies that research, educate, and actively take part in dealing with the global warming issue—in particular, the National Renewable Energy Laboratory (NREL), the National Aeronautics and Space Administration (NASA), the National Oceanic and Atmospheric Administration (NOAA), the Environmental Protection Agency (EPA), and the U.S. Geological Survey (USGS)—for providing an abundance of resources and outreach programs on this important subject. I give special thanks to former vice president Al Gore for his diligent efforts to bring the global warming issue so powerfully to the public's attention. More recently, the California governor Arnold Schwarzenegger has stressed the importance of dealing with global warming in his state. I would especially like to acknowledge the years of leadership and research provided by Dr. James E. Hansen of NASA's Goddard Institute for Space Studies (GISS). His pioneering efforts over the past 20 years have enabled other scientists, researchers, and political leaders worldwide to better understand the scope of the scientific issues involved at a critical point in time when action must be taken before it is too late. I would also like to acknowledge and give thanks to the many wonderful universities across the United States, in England, in Canada, and in Australia, as well as private organizations, such as the World Wildlife Fund, that diligently strive to educate others and help toward finding a solution to this very real problem.

I want to give a huge thanks to my agent, Jodie Rhodes, for her assistance, guidance, and efforts; and also to Frank K. Darmstadt, my editor, for all his hard work, dedication, support, helpful advice, and attention to detail. His efforts in bringing this project to life were invaluable. Thanks also to Alex and the copyediting and production departments for their assistance and the outstanding quality of their work.

INTRODUCTION

One of the biggest human-caused contributors to global warming is the greenhouse gases emitted to the Earth's atmosphere through the continuous burning of fossil fuels. Enormous amounts of greenhouse gases—in order of abundance, water vapor, *carbon dioxide* (CO_2), *methane, nitrous oxide,* and *ozone*—are added daily. For many years, the United States was the largest contributor, but recently China, in the midst of its industrial revolution, has become the world's largest CO_2 emitter at 6,834 million tons (6,200 million metric tons) in 2006 compared to the United States at 6,393 million tons (5,800 million metric tons).

Fossil fuels—oil, natural gas, and coal—are America's primary energy sources, accounting for 85 percent of current U.S. fuel consumption for transportation, industrial, commercial, and residential uses. Among the gases emitted when fossil fuels are burned, one of the most significant is CO_2, a gas that traps heat in the Earth's atmosphere. Over the last 200 years, the burning of fossil fuels has resulted in more than a 25 percent increase in the amount of CO_2 in the atmosphere. Fossil fuels are also implicated in increased levels of atmospheric methane and nitrous oxide, although they are not the major source of these gases.

Black carbon, a form of particulate air pollution most often produced from biomass burning, cooking with solid fuels, and diesel exhaust, has a warming effect on the atmosphere three to four times greater than previously estimated. In fact, soot and other forms of black carbon could contribute as much as 60 percent of the current global warming effect of CO_2, more than any greenhouse gas besides CO_2. In recent years, between 25 to 35 percent of black carbon in the global atmosphere comes from China and India, emitted from the burning of wood and cow dung in household cooking and through the use of coal to heat homes. Countries in Europe and elsewhere that rely heavily on diesel fuel for transportation also contribute large amounts.

Since reliable records began in the late 1800s, the global average surface temperature has risen 0.5–1.1°F (0.3–0.6°C). Scientists with the Intergovernmental Panel on Climate Change (IPCC) concluded in a 1995 report that the observed increase in global average temperature over the last century "is unlikely to be entirely natural in origin" and that the balance of evidence suggests that there is a "discernible human influence on global climate."

Clean air is also essential to life and good health. Several important pollutants are produced by fossil fuel combustion and emitted directly into the atmosphere: carbon monoxide, nitrogen oxides, sulfur oxides, and hydrocarbons. In addition, total suspended particulates (tiny airborne particles of aerosols that are less than 100 micrometers [a micrometer is 1/1000 of a millimeter], which constantly enter the atmosphere from both human [through industrial processes and motor vehicles] and natural [pollen and salt particles] sources) contribute to air pollution, and nitrogen oxides and hydrocarbons can combine in the atmosphere to form tropospheric ozone, the major constituent of smog.

Fossil fuel emissions are added to the atmosphere through several means. The largest contributor is the transportation sector. Cars and trucks are the primary source of carbon monoxide emissions. Two oxides of nitrogen—nitrogen dioxide and nitric oxide—are formed during combustion. Nitrogen oxides appear as yellowish brown clouds over many city skylines. Sulfur oxides are produced by the oxidization of the available sulfur in a fuel. Hydrocarbons are emitted from human-made sources such as auto and truck exhaust. Fossil fuel use also produces particulates, including dust, soot, smoke, and other suspended matter, which are respiratory irritants. Air pollution often forms the prevalent brownish haze that has been termed the *atmospheric brown cloud*. This smog is causing serious environmental effects: It is a public health hazard, causing severe respiratory problems worldwide.

Chapter 1 looks at fossil fuels as energy sources and how much the world population relies on them each day to accomplish tasks. It also outlines the connection of fossil fuels to global warming. Chapter 2 examines the properties of greenhouse gases, the nature and causes of their emissions, who the largest contributors are, and the important health issues and considerations associated with them. Chapter 3 traces the time line from the onset of the Industrial Revolution in the 1700s to the mid-1800s, and it shows how this unique time period changed the world forever, both in development, lifestyle, and the onset of pollution and increase of greenhouse gases in the atmosphere. It also outlines how modernization has had a huge part in the process of global warming and why the newly emerging green industrial revolution is a welcome and essential change.

The next three chapters deal with the specific effects global warming and pollution have on buildings, homes, transportation, cities, and industry—in particular, what energy-efficient homes and commercial buildings must have to ameliorate global warming. Concerning green transportation, the concepts of energy efficiency and fuel economy are discussed. The chief contributors to global warming and pollution on the world's highways are touched on, as well as what new green automobile technology is available. An examination of how cities and industry can make the changes necessary to fight the battle against global warming in order to help the environment in the future and save on energy costs right now follows.

Chapter 7 looks closely at agricultural greenhouse gases and pollution, outlining how this huge industry adds to the global warming issue. It also presents solutions to the problems and provides overviews on how the agricultural sector can also help solve the global warming issue by providing several agricultural resources to produce renewable energy. Finally, this chapter looks at present-day impacts contributing to global warming and practical solutions that have been proposed along with ways everyone can help provide and ensure effective solutions.

Chapters 8 and 9 take a close look at fuel technology and green technology already put in place for tomorrow's cars. Specifically, these chapters address biofuels and clean vehicles, alternative and advanced fuels, and new fuel technology and explain the differences in new car technology including hybrids, electric vehicles, flexible-fuel vehicles, fuel cells, plug-in vehicles, air-powered vehicles, and high-tech cars of the future. The final chapter looks at future energy needs and the direction technology is heading, as well as why public education is key and how each person can contribute in a significant way to solving the problem of global warming.

Energy, Fossil Fuels, and Global Warming

Global warming is the most urgent environmental challenge of the 21st century. Because of the world's continued dependence on *fossil fuels* as an energy source, *greenhouse gas* levels are steadily increasing in the *atmosphere* and warming the Earth. If corrective action is not taken now, temperatures will continue to rise, causing the worldwide destruction of *ecosystem*s and extinction of species. The biggest contributor to warming the atmosphere is the excessive use of fossil fuels for energy. If more efficient technologies are not employed and clean, renewable energy sources such as wind power, solar energy, *fuel cells,* or geothermal energy are not substituted for fossil fuels, there will be no hope of bringing global warming under control. This chapter discusses energy's connection to global warming and why fossil fuels are such major contributors to the problem.

ENERGY SOURCES—FOSSIL FUELS

Fossil fuels are hydrocarbons, derived from coal and petroleum (fuel oil or natural gas). They are formed from the fossilized remains of buried plants and animals that have been subjected to the heat and pressure in the Earth's crust over hundreds of millions of years. Fossil fuels also include substances like oil shale and tar sands, which contain hydrocarbons that are not derived solely from biological sources. These are referred to as mineral fuels.

Today, most of the developed world's industry relies heavily on fossil fuels to produce the energy needed in the manufacture of goods and services. The heat derived from burning fossil fuels is also used for heating and converted to mechanical energy for vehicles and electrical power generation.

We now realize that the burning of fossil fuels is the largest source of carbon dioxide (CO_2) *emissions*. Sadly, their use is steadily increasing. One of the biggest dilemmas we face today is that China, in its race to modernize and industrialize, is building hundreds of coal-fired power plants and increasing CO_2 emissions.

Because fossil fuels are composed almost entirely of *carbon,* when they are burned—such as in a coal-fired plant or in the form of gasoline—the carbon they are composed of is released back into the atmosphere in the form of CO_2. The most common fossil fuels are coal, natural gas, and oil. (Another fossil fuel—liquefied petroleum gas (LPG)—is mainly derived from the production of natural gas.)

Coal

Coal is a sedimentary organic rock that is composed of up to 90 percent carbon by weight. Much of the Earth's coal was formed during the Carboniferous period of the Paleozoic era of geologic time during the Upper, Middle, and Lower Mississippian and Pennsylvanian epochs 354 to 290 million years ago.

As ancient plants and animals died, they were buried in swampy areas and collected in moist peat bogs. Once buried, over time and under enormous heat and pressure, these organic remains were slowly transformed into coal—it took between 4,000 and 100,000 years for three feet (1 m) of peat to accumulate. As these organic deposits were subjected to increasing pressure, the carbon content of the coal became more concentrated. Over time, coal becomes harder and blacker and transforms through a sequence of stages as it ages. The four types of coal are lignite (lowest rank), subbituminous, bituminous, and anthracite (highest rank). The more primitive forms of coal have a lower carbon content and a subsequent lower energy level, whereas the higher in rank, the higher the carbon content, the higher the energy level.

1. **Stage 1: Peat.** Not even a true coal, peat is a fuel used today in several parts of the world. In places where bogs are common (such as Ireland), peat can be cut from the earth, dried out, and burned for its heat value—it is not energy-efficient however.
2. **Stage 2: Lignite.** This coal is soft and brown and similar to peat. It has a low energy output and is composed of about 40 percent carbon. Lignite is not commonly used unless there is nothing else.
3. **Stage 3: Subbituminous.** This coal has an energy content of about 18 million British thermal units (Btu) per ton and is commonly used in coal-fired power plants.
4. **Stage 4: Bituminous.** This is the most common coal used in the United States. Principally from the Carboniferous period about 300 million years ago, it has a high-energy content—about 24 million Btu per ton.
5. **Stage 5: Anthracite.** This is the hardest coal, and is commonly found in Pennsylvania, but the majority of it has already been mined. Containing more than 90 percent carbon, it has a very high energy content—about 23 million Btu per ton—yet unfortunately also has a high sulfur content.

Coal deposits exist in many types of environment. High sulfur coal was formed in saltwater swamps that were once covered by seawater. Coal deposits low in sulfur content were formed under freshwater areas. Seams of coal can be close to the surface of the Earth or buried deep underground. When coal deposits are buried underground, mining for them can be extremely energy intensive and hazardous to miners.

MEASURING ENERGY

Energy is measured in joules, which are very small amounts of energy. A mug of hot chocolate cooling down at room temperature will release about 100,000 joules. The calorie is an old-fashioned unit often used to measure the energy contained in food. A slice of bread contains about 70 calories. One calorie equals about 4,000 joules. Power is the rate at which energy is given off or used, and it is measured in watts. The use of one joule of energy every second is one watt. A 60-watt lightbulb uses 60 joules of energy every second to give off heat and light.

Physical units reflect measures of distance, area, volume, height, weight, mass, force, impulse, and energy. Different types of energy are measured by different physical units: barrels or gallons for petroleum; cubic feet for natural gas; tons for coal; and kilowatt-hours for electricity. To compare different fuels, it is necessary to convert the measurements to the same units. In the United States, the unit of measure most commonly used for comparing fuels is the British *thermal* unit, which is the amount of energy required to raise the temperature of one pound of water one degree Fahrenheit. One Btu is approximately equal to the energy released in the burning of a wood match.

The table opposite illustrates the Btu content of common energy units.

When coal is used in a power plant, pulverized coal is blown into the furnace when it burns. Water flowing through tubes in the furnace is heated to the boiling point under pressure, which blasts through a turbine and turns a generator to produce electricity. After the steam has gone through the turbine it is cooled and *condense*d back into water and returned to the furnace once again, completing a pathway called a Rankine cycle.

When coal burns, it gives off sulfur dioxide, *nitrogen* oxide, CO_2, and other gases. The sulfur particulates are partly removed by scrubbers or filters. Scrubbers use a wet limestone slurry to absorb sulfur. Filters

BTU CONTENT OF COMMON ENERGY UNITS	
1 barrel of crude oil (42 gallons)	5,800,000 Btu
1 gallon of gasoline	124,000 Btu
1 gallon of heating oil	139,000 Btu
1 gallon of diesel fuel	139,000 Btu
1 barrel of residual fuel oil	6,287,000 Btu
1 cubic foot of natural gas	1,026 Btu
1 gallon of propane	91,000 Btu
1 short ton of coal	20,681,000 Btu
1 kilowatt hour of electricity	3,412 Btu

Most people are interested in saving energy these days. The Btu equivalents can be used to compare the pros and cons of using different energy sources. By estimating the total energy usage for each particular energy type, the corresponding Btus can be calculated and compared against each other for a measure of efficiency.

are large cloth bags that capture particles. While up to 90 percent of the sulfur emissions are caught, the remainder can be ejected from the smokestacks into the atmosphere.

Because of its high carbon content, coal emits more CO_2 then any of the other fossil fuels when it is burned. It is also the main source fuel used for the generation of electricity worldwide. According to the Center for Biological Diversity in the United States, an organization that works through science, law, and the media to protect the Earth's lands, water, and *climate* to ensure the survival of species, coal accounts for 83 percent of the greenhouse gas emissions in the electric power sector.

Coal mining is one of the most dangerous types of mining. Underground mining often requires miners to work in close, cramped quarters. *(Department of Energy)*

Globally, coal combustion is the leading contributor to the *anthropogenic* greenhouse gas effect. In addition to CO_2, methane is another by-product of coal. Also a greenhouse gas, methane has a *global warming potential* (GWP) 25 times greater than that of CO_2 over a 100-year span.

Based on a survey conducted by the Union of Concerned Scientists (UCS), the following lists the side effects of a 500-megawatt coal plant that produces 3.5 billion kilowatt-hours each year—enough to power a city of 140,000. Each year it burns 1,430,000 tons of coal, uses 2.2 billion gallons of water, and 146,000 tons of limestone. Each year it releases into the Earth's atmosphere the following:

- 10,000 tons (9,072 metric tons) of sulfur dioxide (SO_x). Sulfur dioxide is the main cause of acid rain, which damages forests, lakes, and buildings.
- 10,200 tons (9,253 metric tons) of nitrogen oxide (NO_x). Nitrogen oxide is a major cause of smog and also a cause of acid rain.
- 3.7 million tons (3.4 million metric tons) of CO_2, the primary greenhouse gas and leading cause of global warming.
- 500 tons (454 metric tons) of small particles. Small particulates are a health hazard, causing lung damage.
- 220 tons (200 metric tons) of hydrocarbons. Fossil fuels are made of hydrocarbons; when they do not burn completely, they are released into the air, causing smog.
- 720 tons (653 metric tons) of carbon monoxide (CO), which is a poisonous gas and contributor to global warming.
- 125,000 tons (113,000 metric tons) of ash and 193,000 tons (175,000 metric tons) of sludge from the smokestack scrubber. The ash and sludge consist of coal ash, limestone, and many pollutants, such as toxic metals like lead and mercury.
- 225 pounds (102 kg) of arsenic, 114 pounds (52 kg) of lead, 4 pounds (2 kg) of cadmium, mercury, trace elements of uranium, and many other toxic heavy metals.

According to the UCS, annual coal production is projected to remain around 2 billion tons (900 million metric tons) into the next century. They estimate that at a steady rate of use, coal *resources* will not be depleted for another 265 years. If, however, the rate of growth continues to increase at 2 percent per year, coal resources will be depleted in 93 years; at a growth rate of 3 percent, it will be depleted in 73 years.

The UCS stresses, however, that even though the physical supplies of coal are plentiful and production costs are currently relatively low, the environmental impacts are enormous. Even though several innovative coal combustion technologies are being developed today and touted as reasonable environmental alternatives, the only practical way to reduce CO_2 emissions from coal is to increase its efficiency by obtaining more energy out of each pound of coal. Today, the efficiency of typical coal plants is only around 33 percent, limited principally by the abilities of the steam turbines. The UCS believes the first way to increase the efficiency of turning coal into electricity is to capture the waste heat in a process called cogeneration. Cogeneration is the generation of heat and power together. It is a well-known but not commonly used technology. One method of cogeneration is to use the waste heat to warm nearby buildings. A process called district heating, while rarely used in the United States, is commonly used in northern Europe.

Another approach with even lower carbon emissions is to run the coal gas through a fuel cell, which converts hydrogen-rich gases, such as methane, into electricity without combustion. Another method currently being researched is a concept called magnetohydrodynamics, or MHD. With MHD, superheated gases from coal combustion blast through a magnetic field created by superconducting magnets, producing an electric charge as they pass. The gases then power a conventional gas turbine, extracting as much energy as possible from the heat. Efficiency levels with this method are estimated to be in the range of 50 to 60 percent.

The UCS cautions, however, that despite several proposed techniques to make the use of coal more efficient and less environmentally damaging, it may never be possible to produce energy from coal without carbon emissions. Most of the heat produced from coal is generated from carbon, which provides more than 70 percent of the energy content. Because there are currently such large reserves of coal in the world and the cost of extracting it is so low, it will take a concerted effort to avoid massive carbon emissions. More efficient use is a good beginning, but replacing coal with renewable energy is the better solution toward the control of global warming.

Natural Gas

Natural gas is also a hydrocarbon, but compared to coal and oil is relatively clean. Natural gas is a product of decomposed organic material. Similar to coal, it is a product of ancient plants and animals that were trapped in bogs and swamps underwater without the presence of oxygen. Over geologic time, these deposits became covered and trapped. Natural gas is formed in porous rock formations, such as sandstone, trapped under a cap of impermeable rock (so that it cannot prematurely escape into the atmosphere). Natural gas is often found mixed with oil or floating on underground reservoirs of oil. The presence of the gas provides the pressure necessary to bring the oil to the surface.

Natural gas deposits are continually being discovered worldwide. According to the UCS, presently known gas reserves in the United States will be able to supply the country for approximately 60 years or more. There are also more natural gas reserves in Russia, Indonesia, Mexico, and North Africa. Estimates of worldwide reserves range from 120 to 175 years of supply. As extraction methods are improved and become more efficient, reserves could be three times higher.

In the early 1900s, natural gas was used largely to light houses and buildings. In the mid-1900s, extensive pipelines were built across the United States. Today there are more than 1 million miles (1.6 million km) of gas lines in the United States. In the United States about half of the natural gas is used by industry, a fourth by the residential sector, and the rest by commercial users and electric utilities. It is used by industry principally for heat, combined heat and electricity (referred to as cogeneration), and as an input for plastics, chemicals, and fertilizer. Gas use in homes is principally for heating purposes.

Although natural gas is a fossil fuel and composed mostly of carbon, emissions that cause global warming are less than those from coal and oil. Natural gas produces 43 percent fewer carbon emissions than coal and 30 percent less than oil. Gas does not produce any solid wastes, such as the large amounts of ash from coal plants. It also produces very little sulfur dioxide (SO_2) and particulate emissions. The combustion of natural gas does, however, produce NO_x and is, in itself, a very powerful greenhouse gas.

Natural gas (methane) is much more effective than CO_2 at trapping heat in the atmosphere—58 times more powerful. Currently, natural gas use has accounted for roughly 10 percent of all global warming emissions.

The use of gas to produce electricity using gas turbines is growing in the United States because gas turbines are cheaper and easier to use than coal plants. Another technology is now becoming available for converting gas to electricity—the fuel cell. Fuel cells convert gas directly into power without combustion. A molecule of natural gas is made up of carbon and hydrogen. When the hydrogen is separated from the carbon and fed into a fuel cell, it combines with oxygen to produce water, electricity, and heat. The carbon is released as carbon dioxide, but in much smaller quantities than from gas turbines.

Fuel cells are highly efficient, converting about 60 percent of the energy in gas into electricity. According to the UCS, as the use of natural gas increases, however, it will become a more important source of greenhouse gases. It will become increasingly important to use it in the cleanest way possible—in fuel cells—in order to reduce global warming impacts.

According to the Center for Biological Diversity, the production of liquefied natural gas (LNG)—natural gas that has been supercooled and converted to liquid for ease of storage or transport—does extreme damage to the climate. It currently takes an enormous amount of energy to liquefy, transport, and regassify LNG. Processing from just one plant can generate more than 24 million tons (22 million metric tons) of greenhouse gases per year, equal to the annual greenhouse gas pollution from about 4.4 million cars. Scientists at Carnegie Mellon University have concluded that LNG can actually produce almost as much greenhouse gas pollution as coal. LNG is manufactured by refrigerating natural gas to condense it into a liquid. Natural gas must be refrigerated to -260°F (-162°C) in order for it to condense. LNG is more than 98 percent pure methane. LPG is a mixture of propane and other similar types of hydrocarbon gases, converted to a liquid state when it is compressed. It must be stored under extreme pressure (~200 pounds per square inch) in order to keep it liquefied.

Oil

Oil has been the key energy source of the 20th century—sometimes referred to as black gold. Worldwide, reserves could supply 40 to 60 years of consumption at present rates. According to the UCS, by the middle of the 21st century, world oil supplies could start dwindling.

Oil, also a hydrocarbon, is formed from the decomposed remains of ancient plants and animals buried under great heat and pressure since the Cambrian period 500 million years ago. The Jurassic (180 to 140 million years ago) and the Cretaceous (140 to 65 million years ago) were also good geologic time periods for oil formation. The organic matter became trapped on ocean floors, mixing with and being covered by sedimentary rocks, such as sandstone and limestone. Similar to natural gas, oil deposits became buried under an impermeable layer of stone or mud. Because the dead plants and animals were trapped under high pressure without oxygen, bacteria were able to break them down into hydrocarbons. Oil is often concentrated in areas where there are high spots (such as geologic rock domes) under the layer of cap rock.

Oil is pumped out of the land with an oil rig. Conflicts arise with conservationists when oil is found on public lands with wilderness, animal habitats, or aesthetic or historical values, and pumping the oil will cause environmental damage. *(Nature's Images)*

Oil must be refined at oil refineries such as this one in Salt Lake City, Utah. *(Nature's Images)*

Once oil has been brought to the surface, it is piped from the oil rig to waiting boats or refineries. Tanker ships worldwide ship crude oil originating from the Middle East. Supertankers—1,312 feet (400 m) in length, the largest moving vehicles ever built—have a cargo capacity of 500,000 tons (454,000 metric tons). The 6,600 existing supertankers in operation worldwide carry 524 billion gallons (2 trillion liters) of oil each year.

In addition to environmental disasters that have occurred over the years from oil spills, air pollution from oil is significant. According to the UCS, transportation accounts for half of NO_x emissions in the United States and a third of CO_2 emissions, as well as emissions of CO, ozone, SO_x, particulates, volatile organic compounds, methane, and toxic metals. These emissions contribute to global warming, urban smog, and acid rain.

Burning petroleum emits about three-fourths as much CO_2 as burning coal and is one of the leading greenhouse gas producers. Besides huge amounts of CO_2 being emitted through the driving of cars, hundreds of millions of tons are emitted in the oil-refining process.

Black Carbon

According to a study featured on the Web site Science*Daily* on March 24, 2008, black carbon, which is a form of particulate air pollution produced by the burning of *biomass,* by cooking with solid fuels, and by diesel exhaust, has a warming effect on the environment three to four times more than what was previously thought. Gregory Carmichael, a chemical engineer at the University of Iowa, and Veerabhadran Ramanathan, an atmospheric physicist at the Center for Clouds, Chemistry and Climate of the Scripps Institution of Oceanography, said that forms of black carbon could have as much as 60 percent of the current global warming effect of CO_2—more than any greenhouse gas besides CO_2. They also believe that the presence of black carbon is playing a role in the present active retreat of arctic sea ice and Himalayan *glacier*s. In their study, Carmichael and Ramanathan used data gathered from *satellite*s, aircraft, and ground instruments and determined that the warming effect of black carbon in the atmosphere is roughly 0.9 watts per meter squared (W/m^2). This estimate is higher than that determined by the Intergovernmental Panel on Climate Change (*IPCC*) in a report issued in 2007 that estimated between 0.2 and 0.9 W/m^2. Carmichael and Ramanathan believe their estimate is more accurate because their computer model *simulation*s do not take into account the amplification of black carbon's warming effect when mixed with sulfates and other *aerosol*s.

Roughly one-fourth of the black carbon content in the atmosphere originates in China and India, where it primarily comes from the burning of wood and cow dung at home for cooking and the use of coal to heat homes. Another major source of black carbon is from countries in Europe that burn diesel for transportation.

According to Ramanathan, "Per capita emissions of black carbon from the United States and some European countries are still comparable to those from both southern and eastern Asia."

In South Asia, where black carbon pollution is especially prevalent, a brownish haze—also called an atmospheric brown cloud—is prominent and is warming the atmospheric temperature and accelerating the melting of Himalayan glaciers that provide drinking water for billions of people throughout Asia.

Not only is black carbon a contributor to global warming, but it is also a public health hazard. There are current technologies available to substantially reduce the levels of black carbon. One positive aspect of this problem is that black carbon particles only remain airborne for a few weeks unlike CO_2, which stays airborne for more than a century.

FOSSIL FUELS AND GLOBAL WARMING

Oil and coal are commonly referred to as fossil fuels. The burning of fossil fuels is one of the leading contributors to global warming. Fossil fuels are made up almost entirely of carbon. In the case of oil, there are other toxic materials that when burned, or when the fumes are inhaled, are known to cause cancer in humans. When coal is burned to generate electricity or oil is burned in the form of gasoline or diesel fuel for transportation, carbon is released into the atmosphere in the form of CO_2.

In developed countries, such as the United States, fossil fuels are the principal sources of energy that are used for fuel, electricity, heat, and air-conditioning. In fact, more than 86 percent of the energy used worldwide originates from fossil fuel combustion. Although for years fossil fuels have been readily available and convenient, they have also played a major role in climate change and global warming. According to the Center for Biological Diversity, fossil fuel use in the United States causes more than 80 percent of the greenhouse gas emissions (Greenhouse gas emissions are discussed in chapter 2.) and 98 percent of just the CO_2 emissions. This adds approximately 4.5 billion tons (4.1 billion metric tons) of CO_2 to the atmosphere each year, which number would be even greater if the Earth did not have natural *carbon sequestration* processes. Nature has provided trees, soil, the oceans, and animals, which act as carbon sinks, or sponges, to soak up the CO_2.

Global warming is under way and will likely continue for the next several centuries due to the long natural processes involved, such as the extended lifetimes of many greenhouse gases. However, there are ways humans can help reduce the potential effects. Because everyone uses energy sources every day, the best way to reduce the negative effects of global warming is to use less energy. By cutting back on the use of electricity, the combustion engine, deforestation, agribusiness, and wasteful lifestyles, fossil fuels can be reduced. The adoption of nonfossil fuel energy sources, such as hydroelectric power, solar power, hydrogen engines, and fuel cells, promises to cut the emission of greenhouse gases in half.

Former vice president Al Gore, the critically acclaimed creator of *An Inconvenient Truth* (2006), the author of *The Assault on Reason* (2007), among many other titles, and the recipient of the 2007 Nobel Peace Prize (received jointly with the IPCC) said, "Climate change is not just a crisis, but the most important crisis mankind has ever faced."

According to a report released by Global Issues, an organization focused on analyzing pressing current scientific, cultural, and political needs, the burning of fossil fuels is creating two separate problems: the greenhouse gases that cause global warming and by-products that are pollutants causing global dimming.

Some of the by-products of fossil fuels such as sulfur dioxide, soot, and ash, are pollutants. When these pollutants enter the atmosphere, they change the properties of the clouds. The pollutants become incorporated into the clouds, resulting in clouds with a larger number of droplets than unpolluted clouds, which makes them more reflective. This causes more of the Sun's heat and energy to be reflected back into space, reducing the heat that reaches the Earth. This phenomenon is called global dimming. In addition to environmental problems, such as smog and acid rain, dimming has also been blamed for contributing to the deaths of millions of people. Because the polluted clouds keep the Sun's heat from reaching the Earth's surface, it has made the waters in the Northern Hemisphere cooler, which has resulted in less rain forming in key areas. Because of this, the Sahel in northern Africa has not received the rainfall it needs. In the 1970s and 1980s, massive famines affected North Africa because of prolonged droughts. According to a 2005 BBC documentary on global dimming by Beate Liepert at Columbia University, when the

High concentrations of particulates trapped in the air in cities are a result of burning fossil fuels—notice how the buildings and mountains in the background are difficult to see. *(Nature's Images)*

data from these decades was run through global dimming models, the computers duplicated the famines experienced in the Sahel. The conclusion was that "what came out of our exhaust pipes and power stations from Europe and North America contributed to the deaths of a million people in Africa, and afflicted 50 million more with hunger and starvation."

According to Anup Shah at Global Issues, the impacts of global dimming might not be in the millions, but billions. The Asian monsoon system is responsible for bringing rainfall to half the world's population. If global dimming affects the Asian monsoons, more than 3 billion people could be negatively affected.

Another contributor to global dimming is the contrails (vapor trails) from airplanes. This was not even understood until the terrorist attacks in the United States on September 11, 2001. Because all commercial flights were grounded for the following three days, it allowed climate scientists to document the effect on the climate when there were no contrails or heat reflection. What they found was that the temperature rose by 1.7°F (1°C) during that three-day period.

Because global dimming has the ability to keep the Earth's temperature slightly cooler, there are concerns that global dimming may in fact be hiding the true power of global warming. *Climate models* today predict a 5°F (3°C) increase in temperature over the next century—a

Jet contrails are another source of global dimming; it was realized after all airline traffic was grounded immediately after the terrorist attacks on September 11, 2001. *(Nature's Images)*

serious increase in temperature. Even grimmer, because global dimming may be masking the full effect of global warming, the temperature increase may be greater than 5°F (3°C).

This could present a conundrum. Global dimming can be alleviated and controlled by cleaning up emissions. If global dimming is addressed but global warming is not, however, the effects of global warming could be amplified. The BBC documentary on dimming suggested that only addressing global dimming would rapidly increase the negative effects of global warming. In that case, it is conceivable that irreversible damage is only about 30 years away. The global impacts could include the melting of ice in Greenland, which would contribute to rising sea levels, inundating coastal locations worldwide; the drying of *tropical* rain forests; and the increase of wildfires, which would release more CO_2 into the atmosphere.

Rather than a 5°F (3°C) temperature increase, there could potentially be a 10°F (6°C) increase. If this were to occur, it would be a more rapid warming than at any other time in history and it would have the following negative effects:

- massive die-offs of vegetation
- a dramatic decrease in food production
- increased soil erosion
- increased *desertification*
- temperature increases
- release of methane hydrate from the oceans' bottoms—a greenhouse gas eight times stronger than CO_2

According to the BBC documentary, "This is not a prediction, it is a warning of what will happen if we clean up the pollution while doing nothing about greenhouse gases."

Using global dimming as a way to curtail the effects of global warming, however, is not a viable option. Allowing pollutants to remain in the atmosphere will cause health problems from soot and smog such as respiratory illnesses. It will also lead to increased environmental problems such as acid rain, as well as *ecological* problems such as changes in rainfall patterns, which can lead to the deaths of millions of people from drought and failed agriculture systems. Instead of dealing with just one or the other, both global warming and global dimming must be dealt with together.

The IPCC consensus in their 2007 report is that "Humankind's reliance on fossil fuels—coal, fuel oil, and natural gas—is to blame for global warming." According to a report in *USA Today*, Jerry Mahlman, an American *climatologist* who was formerly the director of the federal Geophysical Fluid Dynamics Laboratory in New Jersey, said, "The IPCC report represents a real convergence, a consensus that this [global warming] is a total global no-brainer."

Claudia Tebaldi, a researcher at the National Center for Atmospheric Research (NCAR) in Boulder, Colorado, remarked, "The big message that will come out is the strength of the attribution of the warming to human activities." Commenting about the IPCC's 2007 report, she said, "The report lays blame at the feet of fossil fuels with 'virtual certainty,' meaning 99 percent sure."

This is a significant increase from their 2001 report, where they said fossil fuels were likely, or 66 percent sure. The upgraded assessment was the result of a two-month intensive review of more than 1,600 pages of new research data compiled by more than 2,500 scientists.

According to Tabaldi, "Even if people stop burning the fossil fuels that release carbon dioxide, the heat trapping gas blamed most for the warm up, the effects of higher temperatures, including deadlier heat waves, coastal floods, longer droughts, worse wildfires, and higher energy bills, would not go away in our lifetime.

"The projections also make clear how much we are already committed to climate change. Even if every smokestack and tailpipe stops emissions right now, the remaining heat makes further warming inevitable."

Mahlman added, "Most of the carbon dioxide still would just be sitting there, staring at us for the next century."

The 2007 report also pointed out that this issue should finally receive the serious attention it deserves. Previously, the argument was focused more on whether the problem was natural or human-induced.

In the two years since its release, there have been several achievements and advancements made. The levels of research have grown, public awareness has increased, the subject has been incorporated into many school curriculums worldwide, and legislation—local, national, and international—has been passed and is currently being introduced into governments around the world. In addition to the 2007 Nobel Peace Prize shared by Al Gore and the IPCC, the film *An Inconvenient Truth* earned two Oscars at the 2007 Academy Awards ceremony in Los Angeles, California. Global warming issues are finally receiving the media attention they deserve, making the public more aware of the real issues and the reasons why they need to be addressed now.

Growing public education and awareness have not solved all the problems. Although the public is becoming more educated, skeptics are also raising their voices in protest, continuing to cloud the issues, making it important for people to pay attention to the facts. Many cities worldwide, foreign countries, and individual states in the United States are taking action to curb fossil fuel emissions. Arnold Schwarzenegger, governor of California, for example, has ordered the world's first low-carbon limits on passenger car fuels in the most populous state. The new standard reduces the carbon content of transportation fuels at least 10 percent by the year 2020.

Climatologists at Lawrence Livermore National Laboratory in California created a climate and *carbon cycle* model to examine global climate and carbon cycle changes. What they concluded was that if humans continued with the same lifestyles and habits they are accustomed to today (commonly referred to as a business-as-usual approach), the Earth's atmosphere would warm by 14.5°F (8°C) if humans use all of the Earth's available fossil fuels by the year 2300.

Their model predicted several alarming results: In the next few centuries, the polar ice caps will have vanished, ocean levels will rise by 23 feet (7 m), in the polar regions temperatures will climb higher than the average predicted 14.5°F (8°C) to 33°F (20°C), transforming the delicate ecosystems from polar and *tundra* to boreal forest. Govindasamy Bala, of the Lawrence Livermore National Laboratory's Energy and Environment Directorate and lead author of the project, said, "The temperature estimate was actually conservative because the model didn't take into consideration changing *land use* such as *deforestation* and build-out of cities into outlying wilderness areas."

While current atmospheric CO_2 levels are 380 *parts per million (ppm)*, the model projected that by 2300 the level will have risen to 1,423 ppm—a nearly 400 percent increase. The model identified the soil and biomass as significant carbon sinks. But, according to Bala, "The land ecosystem would not take up as much carbon dioxide as the model assumes. In fact, in the model, it takes up much more carbon than it would in the real world because the model did not have nitrogen/nutrient limitations to uptake. We also didn't take into account *land use changes.*"

The results of the model showed that ocean uptake of CO_2 starts to decrease in the 22nd and 23rd centuries as the ocean surface warms. It takes longer for the ocean to absorb CO_2 than it does for the vegetation and the soil. By 2300, the land will absorb 38 percent of the CO_2 released from the burning of fossil fuels, and 17 percent will be absorbed by the oceans. The remaining 45 percent will stay in the atmosphere. Over time, roughly 80 percent of all CO_2 will end up in the oceans via physical processes, increasing its acidity and harming aquatic life.

Ken Caldeira of the Department of Global Ecology at the Carnegie Institution and another author of the project, said, "The doubled CO_2 climate that scientists have warned about for decades is beginning to look like a goal we might attain if we work hard to limit CO_2 emissions, rather than the terrible outcome that might occur if we do nothing."

According to Bala, the most drastic changes during the 300-year period will occur during the 22nd century—when precipitation patterns change, when an increase in the amount of atmospheric precipitable water and a decrease in the size of sea ice are the largest, and when emission rates are the highest. Based on the model's results, all sea ice in the Northern Hemisphere summer will have vanished by 2150.

When referring to global warming skeptics, Bala says, "Even if people don't believe in it today, the evidence will be there in 20 years. These are long-term problems. We definitely know we are going to warm over the next 300 years. In reality, we may be worse off than we predict."

GLOBAL ENERGY USE

The energy required to support individual lifestyles, industries, and economies provides conveniences today unlike any civilizations have ever been able to enjoy in the past. But for all of the resulting benefits and comforts enjoyed today because of these energy sources, enormous costs on human health and the Earth's natural ecosystems are being exacted. As populations continue to grow, developing countries continue to become more advanced and industrialized, and developed countries grow and advance technologically, enormous amounts of energy are required.

Even though there have been technological advancements in the past few decades in *energy efficiency,* the level of worldwide energy consumption has continued to increase. According to the Worldwatch Institute, between 1850 and 1970 the number of people living on Earth more than tripled, but the energy they consumed rose more than 12-fold. By 2002, the world population had grown another 68 percent and fossil fuel consumption had risen another 73 percent.

In the past decade, U.S. oil use has increased almost 2.7 million barrels a day. For a comparison, the average American consumes five times more energy than the average person worldwide. Energy consumption is currently rising fastest in the developing countries, where petroleum use has quadrupled since 1970. China holds the place as the world's largest coal consumer and the third-largest oil user. Global energy usage must be cut back; it is not sustainable at the present rate. In the future, for example, if the average Chinese consumer used as much oil as the average American uses today, China would require 90 million barrels per day—11 million more than the entire world produced each day in 2001.

Transportation accounts for 30 percent of the world's energy use and is the world's fastest-growing form of energy use. It gobbles 95 percent of global oil consumption. For a dramatic picture, in 2002, 40.6 million passenger cars were manufactured, an amount five times greater than what was manufactured in 1950. Today, there are more than 531 million cars worldwide, and that number is growing by approximately 11 million cars each year—and one-fourth of them are in the United States, where cars and light trucks are responsible for roughly 40 percent of the nation's oil use and contribute about as much to climate change as all the economic activity in Japan.

Worldwide, roughly one-third of all energy produced is used in buildings for heating, cooling, lighting, cooking, and running appliances, equipment, and machines. Currently, building-related energy demand is rising rapidly—especially in private homes. Energy use in homes varies nationally. Populations in the United States and Canada use 2.4 times as much energy at home as those in western Europe.

Home sizes are growing as well. In the United States alone, the average size of new homes grew almost 38 percent between 1975 and 2000, to twice the size of typical homes in Europe and Japan and 26 times larger than those of inhabitants in Africa. As houses become larger, they require more and more energy to heat, cool, light, and run appliances. Home appliances are the world's fastest-growing energy consumers

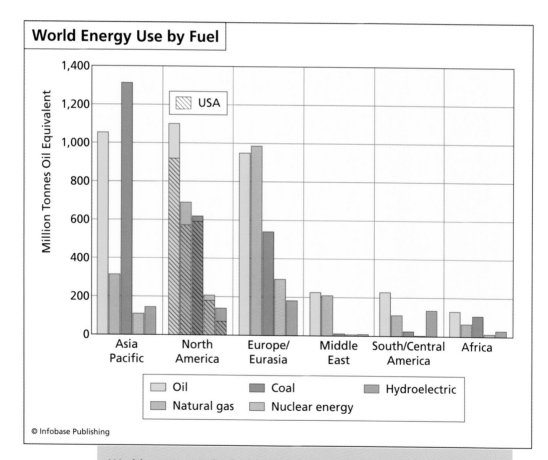

World energy use by fuel type. The chart includes major commercially traded fuels only. (Source: BP Statistical Review of World Energy, 2004)

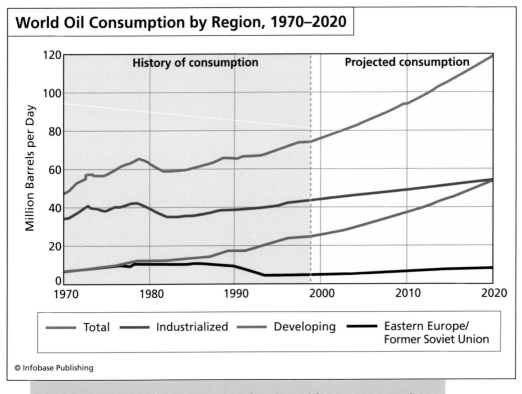

World oil consumption by region and projected future consumption
(Source: Association for the Study of Peak Oil)

after cars and account for 30 percent of industrial countries' electricity consumption and 12 percent of the greenhouse gas emissions.

Every commodity purchased by a consumer has associated energy inputs related to it. The largest share of global energy consumption goes into producing vehicles, buildings, appliances, clothes, and food. According to the Worldwatch Institute, "People can live in a typical house for ten years before the energy they use in it exceeds what went into the manufacture of its components (steel beams, cement foundation, windows, drywall, tile floors, carpet, and its construction)."

Manufacturing plants also place a large demand on energy and produce resultant greenhouse gases, increasing the negative effects of global warming. The production of food also requires substantial amounts of energy. In fact, nearly 21 percent of the fossil energy consumed each year is used on the global food system. David Pimentel, professor of ecology and agriculture at Cornell University, estimates that the United States uses about 17 percent of its fossil fuel consumption for the production and consumption of food.

In the IPCC's third report on global warming in 2001, they stressed that most of global warming and its resultant havoc wreaked on the world's ecosystems can be blamed on human activity.

William C. G. Burns, writing for California's Pacific Institute for Studies in Development, Environment, and Security, said, "The IPCC report indicates that a 60 to 70 percent reduction in greenhouse gas emissions is needed in order to stabilize the land of pollutants in the air—an amount that has doubled since the beginning of the *industrial revolution*. Long- and short-term effects of global warming will worsen as greenhouse gases are added to the atmosphere and could have horrific implications. Warming may eventually change marine and terrestrial systems that humans depend on. One model predicts that 22nd-century Europe will become an arctic environment because 'the system becomes overwhelmed.' That's by no means believed to be science fiction."

A significant way to prevent global warming from creating a situation like this is to cut back immediately on the use of fossil fuels. It must, however, be a global effort in order to be effective.

Greenhouse Gases, Health, and the Environment

Greenhouse gases are generated throughout the world. Specific activities and sources are responsible for certain gases. Each different type of gas has its own unique characteristics, such as potency and life span. Gaining an understanding of these differences and their implications for the environment is critical in the management and understanding of global warming. This chapter looks at the role of greenhouse gases pertaining to global energy use. It then explores their long-term effects and who are the biggest contributors to this problem. Health issues associated with greenhouse gases and global warming are also examined. These are the issues that the leaders of tomorrow need to start thinking about today.

GREENHOUSE GASES

Trace gases in the atmosphere act like glass in a greenhouse. The trace gases serve to trap the heat energy from the Sun close to Earth. Most greenhouse gases occur naturally and are cycled through the global biogeochemical system (the continual interaction between the hydrosphere [water], biosphere [life-forms], lithosphere [landforms], and atmosphere [air]). However, it is the greenhouse gases being added by human activity that are trapping an excessive amount of heat today and causing the atmosphere to overheat. There are several different types of greenhouse gases, and some exist in greater quantities than others. They include water vapor, carbon dioxide (CO_2), methane (CH_4), nitrous oxide (N_2O), and halocarbons. Greenhouse gases capture 70 to 85 percent of the energy in up-going thermal *radiation* emitted from the Earth's surface.

Water vapor is the most common greenhouse gas—it accounts for roughly 65 percent of the *greenhouse effect*. When water heats up, it evaporates into vapor and rises from the Earth's surface into the atmosphere, forming clouds that act as either an insulating blanket to keep the Earth warm or reflect and scatter incoming sunlight. This is why cloudy nights are warmer than clear nights. As water vapor condenses and cools, it comes back to Earth as snow or rain and continues on its way through the water cycle.

CO_2 is generated from several sources. The second most prevalent greenhouse gas, it accounts for about 25 percent of the greenhouse effect. Humans and animals exhale CO_2, vegetation releases CO_2 when it dies and decomposes, burning trees in a forest fire or during deforestation releases CO_2, and burning fossil fuels (such as exhaust from cars and industrial processes) create a huge amount of CO_2.

Methane is a colorless, odorless, flammable gas, formed when plants decay in an environment with very little air. It is the third most common greenhouse gas and is created when organic matter decomposes without the presence of oxygen—a process called anaerobic decomposition. One of the most common sources of methane is from ruminants—grazing animals with multiple stomachs to digest their food. These include cattle, sheep, goats, camels, bison, and musk ox. In their digestive systems, their large fore-stomach hosts tiny microbes that break down their food. This process creates methane gas, which is released as flatulence or belching. In fact, in one day a single cow can emit one-half pound of methane into the air. Each day, 1.3 billion cattle burp methane several times per minute.

Methane is also a by-product of natural gas and decomposing organic matter, such as food and vegetation. Present in wetlands, it is commonly referred to as swamp gas. Since 1750, methane has doubled its *concentration* in the atmosphere and is projected to double again by 2050. According to Nick Hopwood and Jordan Cohen at the University of Michigan, every year 350 to 500 million tons of methane are added to the atmosphere through various activities, such as the raising of livestock, coal mining, drilling for oil and natural gas, garbage sitting in landfills, and rice cultivation. Rice cultivation is a huge global business. In the past 50 years, rice farmland has doubled in area. Rice is a major food staple; it currently feeds one-third of the world's population. Because rice is grown in waterlogged soils, like swamps, it releases methane as a by-product.

Nitrous oxide is released from manure and nitrogen-based chemical fertilizers. As the fertilizer breaks down, N_2O is released into the atmosphere. Also, N_2O is contained in soil by bacteria. When farmers plow the soil and disturb the surface layer, N_2O is released into the atmosphere. It is also released from catalytic converters in cars and from the ocean. According to Hopwood and Cohen, N_2O has risen more than 15 percent since 1750. Each year 7 to 13 million tons (6 to 12 million metric tons) is added to the atmosphere principally through the use of nitrogen-based fertilizers, the disposal of human and animal waste in sewage treatment plants, automobile exhaust, and other sources that have not been identified yet. The use of nitrogen-based fertilizers has doubled in the last 15 years. Although good for the productivity of crops, they are not good for the atmosphere.

Halocarbons include the fluorocarbons, methylhalides, carbon tetrachloride, carbon tetrafluoride, and halons. These are all powerful greenhouse gases because they strongly absorb terrestrial infrared radiation and stay in the atmosphere for many decades.

Fluorocarbons are a group of synthetic organic compounds that contain fluorine and carbon. One of the common compounds is chlorofluorocarbon (CFC). This class contains chlorine atoms and has been used in industry as refrigerants, cleaning solvents, and propellants in spray cans. These CFCs are harmful to the atmosphere because they deplete the ozone layer, and their use has been banned in most areas of the world, including the United States.

Hydrofluorocarbons (HFCs) contain fluorine and do not damage the ozone layer. Fluorocarbon polymers are chemically inert and electrically insulating. They are used in place of CFCs because they do not harm or break down ozone molecules, but they do trap heat in the atmosphere. HFCs are used in air-conditioning and refrigerators. The best way to keep them out of the atmosphere is to recycle the coolant from the equipment they are used in.

Fluorocarbons have several practical uses. They are used in anesthetics in surgery, as coolants in refrigerators, as industrial solvents, as lubricants, as water and stain repellants, and as chemical reagents. They are used to manufacture fishing line and are contained in products such as Gore-Tex and Teflon.

GLOBAL WARMING POTENTIAL

The different types of greenhouse gases all have different properties. For example, the amount of time they reside in the atmosphere and the amount of heat that they trap can vary widely.

Many of the greenhouse gases are extremely potent—some can continue to reside in the atmosphere for thousands of years after they have been emitted. According to the U.S. Environmental Protection Agency (EPA), some greenhouse gases are 140 to 23,900 times more potent than CO_2 in terms of their ability to trap and hold heat in the atmosphere over a 100-year period. It is important to note that these gases and their effects will continue to increase in the atmosphere as long as they continue to be emitted. Even though these gases represent a very small proportion of the atmosphere—less than 2 percent of the total—their enormous heat-holding potential makes them significant and represents a serious addition to global warming.

In order to understand specific greenhouse gases' potential impact, they are rated as to their global warming potential (GWP). The GWP of a greenhouse gas is the ratio of global warming—or radiative *forcing*—from one unit mass of a greenhouse gas to that of one unit mass of CO_2 over a period of time, making the GWP a measure of the "potential for global warming per unit mass relative to carbon dioxide." In other words, greenhouse gases are rated on how potent they are compared to CO_2.

GWPs take into account the absorption strength of a molecule and its atmospheric lifetime. Therefore, if methane has a GWP of 23 and carbon has a GWP of 1 (the standard), this means that methane is 23 times more powerful than CO_2 as a greenhouse gas. The Intergovernment Panel on Climate Change (IPCC) has published reference values for GWPs of several greenhouse gases. Reference standards are also issued and supported by the United Nations Framework Convention on Climate Change (UNFCCC), as shown in the table on the next page.

The higher the GWP value, the larger the infrared absorption and the longer the atmospheric lifetime. Based on this table, even small amounts of sulfur hexafluoride and HFC-23 can contribute a significant amount to global warming.

The EPA has identified three major groups of high GWP gases: (1) hydrofluorocarbons, (2) perfluorocarbons (PFCs), and (3) sulfur hexafluoride. These represent the most potent greenhouse gases. Also, PFCs and sulfur hexafluoride have extremely long atmospheric lifetimes—up to 23,900 years. Since their lifetimes are so incredibly long, for practical purposes once they are emitted into the atmosphere, they are considered to be there permanently. According to the EPA, once they are present in the atmosphere, the result is "an essentially irreversible accumulation."

Hydrofluorocarbons are man-made chemicals. Most of them were developed as replacements for the ozone-depleting substances that were common in industrial, commercial, and consumer products. The GWP index for HFCs ranges from 140 to 11,700. Their lifetime in the atmosphere ranges from one to 260 years; the most commonly used ones have a lifetime of about 15 years and are used in automobile air-conditioning and refrigeration.

Perfluorocarbons generally originate from the production of aluminum and semiconductors. PFCs have very stable molecular structures and usually do not get broken down in the lower atmosphere. When they reach the mesosphere 37 miles (60 km) above the Earth's surface, high-energy ultraviolet electromagnetic energy destroys them, but it is

Global Warming Potential of Greenhouse Gases		
GREENHOUSE GAS	LIFETIME IN THE ATMOSPHERE	GWP OVER 100 YEARS (COMPARED TO CO_2)
carbon dioxide	50–200 years	1
methane	12 years	23
nitrous oxide	120 years	296
CFC 115	550 years	7,000
HFC-23	264 years	11,700
HFC-32	5.6 years	650
HFC-41	3.7 years	150
HFC-43-10mee	17.1 years	1,300
HFC-125	32.6 years	2,800
HFC-134	10.6 years	1,000
HFC-134a	14.6 years	1,300
HFC-152a	1.5 years	140
HFC-143	3.8 years	300
HFC-143a	48.3 years	3,800
HFC-227ea	36.5 years	2,900
HFC-236fa	209 years	6,300
HFC-245ca	6.6 years	560
sulfur hexafluoride	3,200 years	23,900
perfluoromethane	50,000 years	6,500
perfluoroethane	10,000 years	9,200
perfluoropropane	2,600 years	7,000
perfluorobutane	2,600 years	7,000
perfluorocyclobutane	3,200 years	8,700
perfluoropentane	4,100 years	7,500
perfluorohexane	3,200 years	7,400

Source: UNFCCC

a very slow process, which enables them to accumulate in the atmosphere for several thousand (up to 50,000) years.

Sulfur hexafluoride has a GWP of 23,900, making it the most potent greenhouse gas. It is used in insulation, electric power transmission equipment, the magnesium industry, semiconductor manufacturing to create circuitry patterns on silicon wafers, and as a tracer gas for leak detection. Its accumulation in the atmosphere shows the global average concentration has increased by 7 percent per year during the 1980s and 1990s—according to the IPCC, from less than one part per trillion (ppt) in 1980 to almost four ppt in the late 1990s.

In response to global warming, the EPA is working to reduce the emission of gases with high GWP because of their potency and long lifetimes. Major emission sources of high-GWP gases are from industries such as electric power generation, magnesium production, semiconductor manufacturing, and aluminum production.

In electric power generation, sulfur hexafluoride is used in circuit breakers, gas-insulated substations, and switchgear. During magnesium metal production and casting, sulfur hexafluoride serves as a protective cover gas during processing. It improves safety and metal quality by preventing the oxidation and potential burning of molten magnesium in the presence of air. It replaced sulfur dioxide, which was more environmentally toxic. The semiconductor industry uses many high GWP gases in plasma etching and in cleaning chemical vapor deposition tool chambers. They are used to create circuitry patterns. During primary aluminum production, GWP gases are emitted as by-products of the smelting process.

The best solution found to date to ameliorate the negative impact on the environment and to combat global warming is through the EPA working with private industry to develop and implement new processes that are better for the environment. In addition, the EPA is working to limit high GWP gases through mandatory recycling programs and restrictions.

If a greenhouse gas can remain in the atmosphere for several hundred years, even though it may be in a small amount, it can do a substantial amount of damage. Some of the greenhouse effect today is due to greenhouse gases put in the atmosphere decades ago. Even trace amounts can add up significantly.

GREENHOUSE EMISSIONS AND THE BIGGEST CONTRIBUTORS

Carbon dioxide enters the air during the carbon cycle and comes from several sources. Vast amounts of carbon are stored naturally in the Earth's soils, oceans, and sediments at the bottoms of oceans. Carbon is stored in the Earth's rocks and is released when the rocks erode. Carbon exists in all living matter. Every time animals and plants breathe, they exhale CO_2.

The Earth maintains a natural carbon balance. Throughout geologic time, when concentrations of CO_2 have been disturbed, the system had always gradually returned to its natural (balanced) state. This natural readjustment works very slowly.

Through a process called diffusion, various gases that contain carbon move between the ocean's surface and the atmosphere. Because of this, plants in the ocean use CO_2 from the water for *photosynthesis,* which means that ocean plants store carbon, just as land plants do. When ocean animals eat these plants, they then store the carbon. Then, when they die and decompose, they sink to the bottom, and their remains become incorporated in the sediments on the ocean bottom.

Once in the ocean, the carbon can go through various processes: it can form rocks and *weather,* and it can be used in the formation of shells. Carbon can move to and from different depths of the ocean and also exchange with the atmosphere.

As carbon moves through the system, different components can move at different speeds. Scientists break these reaction times down into two categories: short-term cycles and long-term cycles. In short-term cycles, carbon is exchanged quickly. Examples of this include the gas exchange between the oceans and atmosphere (*evaporation*). Long-term cycles can take anywhere from years to millions of years to occur. Examples of this are carbon stored for years in trees or weathered carbon from a rock being carried to an ocean, buried and incorporated into plate tectonic systems, then later being released into the atmosphere through a volcanic eruption.

Throughout geologic time, the Earth has been able to maintain a balanced carbon cycle. The danger today is that this natural balance has been upset by recent human activity. Over the past 150 to 200 years, fossil fuel emissions, land-use changes, and other human activities have increased atmospheric CO_2 by 30 percent (and methane by 150 percent) to concentrations not seen in the past 420,000 years.

Humans are adding CO_2 to the atmosphere much faster than the Earth's natural system can remove it. Prior to the Industrial Revolution, atmospheric carbon levels remained constant at around 280 parts per million (ppm). This meant that the natural carbon sinks were balanced between what was being emitted and what was being stored. After the industrial revolution began and CO_2 levels began to increase—315 ppm in 1958 to 383 ppm in 2007—the balancing act became unbalanced, and the natural sinks could no longer store as much carbon as was being introduced into the atmosphere by human activities. In addition, according to Dr. Pep Canadell of the National Academy of Sciences, 50 years ago, for every ton of CO_2 emitted, 1,323 pounds (600 kg) were removed by natural sinks. In 2006, only 1,213 pounds (550 kg) were removed per ton, and the amount continues to fall today, which indicates that the natural sinks are losing their carbon-storage efficiency. This means that while the world's oceans and land plants are absorbing great amounts of carbon, they cannot keep up with what humans are adding. The natural processes work much more slowly than the human ones do. The Earth's natural cycling usually takes millions of years to move large amounts from one system to another. The problem with human interference is that drastic changes to the environment are happening in only centuries or decades—and the Earth cannot keep up with the fast pace. The result is that each year the measured CO_2 concentration of the atmosphere gets higher, making the Earth's atmosphere warmer.

Levels of several greenhouse gases have increased by about 25 percent since large-scale industrialization began about 150 years ago. According to the National Energy Information Center (NEIC) of the Energy Information Administration (EIA), 75 percent of the anthropogenic CO_2 emissions added to the atmosphere over the past 20 years are due to the burning of fossil fuels.

According to the EIA, natural Earth processes can absorb approximately 3.5 billion tons (3.2 billion metric tons) of anthropogenic CO_2 emissions annually. An estimated 6.7 billion tons (6.1 billion metric tons) is added each year, however, creating a large imbalance, which is why there is a steady, continuous growth of greenhouse gases in the atmosphere.

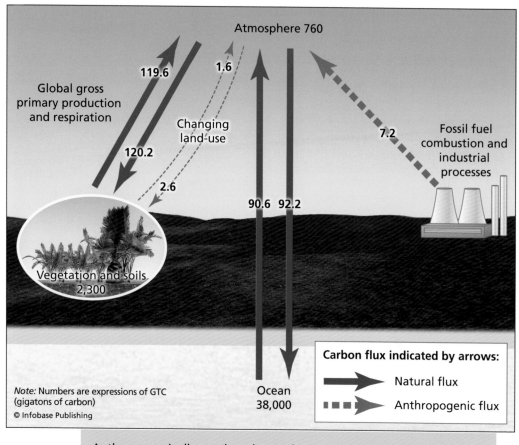

Atmosphere 760

119.6   1.6

Global gross
primary production
and respiration

Changing
land-use

120.2

Fossil fuel
combustion and
industrial
processes

7.2

2.6

90.6 92.2

Vegetation and soils
2,300

Ocean
38,000

Note: Numbers are expressions of GTC
(gigatons of carbon)
© Infobase Publishing

Carbon flux indicated by arrows:

Natural flux

Anthropogenic flux

Anthropogenically produced greenhouse gases contribute much
more CO_2 to the Earth's atmosphere than natural processes do.
(Source: Intergovernmental Panel on Climate Change)

In computer models, an increase in greenhouse gases results in
an increase in average temperatures on Earth. The warming that has
occurred over the past century is largely attributed to human activity.
According to a study conducted by the National Research Council in
May, 2001: "Greenhouse gases are accumulating in Earth's atmosphere
as a result of human activities, causing surface air temperatures and
sub-surface ocean temperatures to rise. Temperatures are, in fact, rising. The changes observed over the last several decades are likely mostly
due to human activities."

The following table shows the U.S. energy-related CO_2 emissions by
fossil fuel type.

U.S. Energy-Related CO_2 Emissions by Fossil Fuel Type (million metric tons CO_2)

YEAR	PETROLEUM	COAL	NATURAL GAS	TOTAL
1995	2,206	1,893	1,192	5,301
1996	2,287	1,976	1,215	5,489
1997	2,309	2,025	1,225	5,570
1998	2,352	2,045	1,198	5,607
1999	2,414	2,046	1,198	5,669
2000	2,458	2,140	1,239	5,848
2001	2,469	2,084	1,190	5,754
2002	2,468	2,093	1,245	5,820
2003	2,513	2,130	1,216	5,872
2004	2,604	2,155	1,196	5,966
2005	2,621	2,163	1,179	5,974
2006	2,586	2,132	1,158	5,888
2007	2,583	2,154	1,234	5,984
2008	2,413	2,130	1,247	5,802

* According to the EIA, energy-related CO_2 emissions declined by 2.8 percent in 2008 due to higher energy prices and lower economic growth. Total energy consumption in 2008 fell by 2.2 percent.
Source: Energy Information Administration

The EIA has determined that greenhouse gas emissions originate principally from energy use, driven mainly by economic growth, fuel used for electricity generation, and weather patterns affecting heating and cooling needs. Energy-related CO_2 emissions, resulting from petroleum and natural gas, represent 82 percent of the total U.S. human-made greenhouse gas emissions. According to the EIA, the graph represents

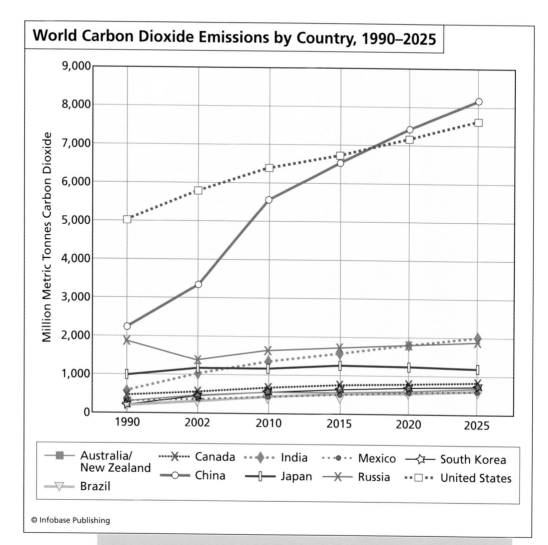

World Carbon Dioxide Emissions by Country, 1990–2025

Legend:
- Australia/New Zealand
- Brazil
- Canada
- China
- India
- Japan
- Mexico
- Russia
- South Korea
- United States

© Infobase Publishing

This is a ranking of the world's countries according to CO_2 emissions due to the burning of fossil fuels in thousands of metric tonnes per annum. The United States and China (dark blue) are the highest emitters.

historical and projected world CO_2 emissions by the top CO_2-emitting countries from 1990 to 2025.

The tables on pages 39 and 40 illustrate CO_2 emissions from the use of fossil fuels by the world's top 20 emitters, as well as major worldwide geographic regions, as of 2006 based on data collected from the EIA and the Oak Ridge National Research Laboratory.

(continues on page 42)

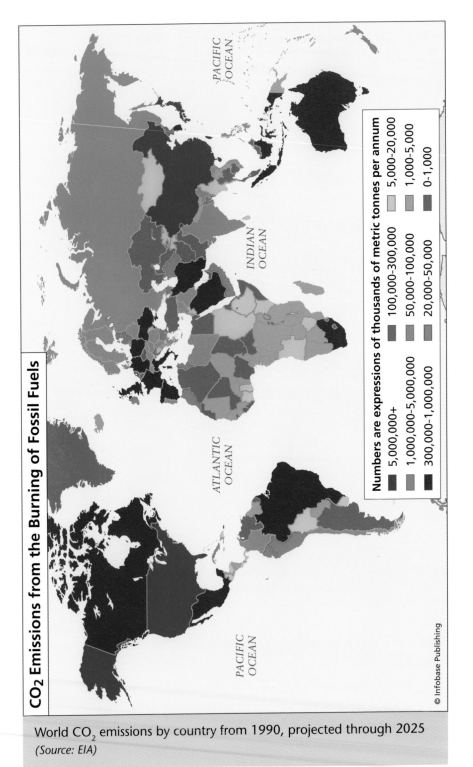

CO₂ Emissions from the Burning of Fossil Fuels

Numbers are expressions of thousands of metric tonnes per annum

5,000,000+
1,000,000-5,000,000
300,000-1,000,000
100,000-300,000
50,000-100,000
20,000-50,000
5,000-20,000
1,000-5,000
0-1,000

PACIFIC OCEAN
INDIAN OCEAN
ATLANTIC OCEAN
PACIFIC OCEAN

© Infobase Publishing

World CO₂ emissions by country from 1990, projected through 2025
(Source: EIA)

The Top 20 CO$_2$ Emitters	
COUNTRY	CO$_2$ EMISSIONS (MILLION METRIC TONS OF CO$_2$)
China (mainland)	6,017.69
United States	5,902.75
Russian Federation	1,704.36
India	1,293.17
Japan	1,246.76
Germany	857.60
Canada	614.33
United Kingdom	585.71
Republic of Korea	514.53
Iran	471.48
Italy (including San Marino)	468.19
South Africa	443.58
Mexico	435.60
Saudi Arabia	424.08
France (including Monaco)	417.75
Australia	417.06
Brazil	377.24
Spain	372.62
Ukraine	328.72
Indonesia	280.36

World CO$_2$ Emissions (million metric tons) by Geographic Region, 2001–2006, 2030

GEOGRAPHIC REGION	2001	2002	2003	2004	2005	2006	PROJECTED FOR 2030
North America	6,697.34	6,782.00	6,870.49	6,970.01	7,034.15	6,954.03	7,703.00
Central and South America	1,015.58	1,005.01	1,022.68	1,065.71	1,110.78	1,138.49	1,654.00
Europe	4,559.17	4,532.33	4,678.65	4,713.13	4,717.46	4,720.85	5,897.00
Eurasia	2,332.38	2,354.20	2,470.57	2,528.65	2,599.84	2,600.65	3,422.00
Middle East	1,118.75	1,175.37	1,240.40	1,330.10	1,444.16	1,505.30	2,279.00
Africa	922.55	924.10	974.71	1,024.82	1,061.61	1,056.55	1,409.00
Asia and Oceania	7,607.73	8,050.28	8,806.46	9,820.89	10,517.00	11,219.56	15,917.00
World Total	**24,253.49**	**24,823.30**	**26,063.96**	**27,453.30**	**28,485.00**	**29,195.42**	**38,281.00**

THE STATE OF THE UNION IN TOTAL GREENHOUSE GAS EMISSIONS

According to the EIA's annual report, released on December 3, 2008, the status of greenhouse gas emissions in the United States includes the following points:

- Total U.S. greenhouse gas emissions in 2007 were 1.4 percent above 2006 levels.
- Total emissions growth—from 7,179.7 million metric tons of CO_2 equivalent in 2006 to 7,282.4 million in 2007—was largely the result of an increase in CO_2 emissions. There were larger percentage increases in emissions of other greenhouse gases (methane, N_2O, and man-made gases with high GWPs), but their absolute contributions to total emissions growth were relatively small.
- The increase in U.S. CO_2 emissions in 2007 resulted primarily from two factors—unfavorable weather conditions, which increased demand for heating and cooling in buildings, and a drop in hydropower availability that led to greater reliance on fossil energy sources (coal and natural gas) for electricity generation, increasing the carbon intensity of the power supply.
- An economic growth of 2.0 percent in 2007, coupled with a 1.4 percent increase in total greenhouse gas emissions, accounted for the lack of improvement in U.S. greenhouse gas intensity from 2006 to 2007.
- Since 2002, the base year for the Bush administration's emissions intensity reduction goal of 18 percent in a decade, U.S. greenhouse gas intensity has been reduced 9.8 percent from 2002 to 2007.
- The steady increase in carbon intensity has resulted mainly from reductions in energy use rather than increased use of low-carbon fuels.
- In the EIA's *International Energy Outlook 2008,* the U.S. share of world CO_2 emissions is projected to fall to 16 percent in 2030.

(continues)

(continued)

- The Consolidated Appropriations Act of 2008, which became Public Law 110-161 on December 26, 2007, directed the EPA to develop a draft mandatory reporting rule for greenhouse gases by the end of September 2008. Although the draft rule has not yet been released, the final rule is due to be completed by June 2009. It is expected to require mandatory reporting of greenhouse gas emissions "above appropriate thresholds in all sectors of the economy," with thresholds and frequency of reporting to be determined by the EPA.

Included in the EIA's annual report are data on energy consumption and production; overviews of petroleum, natural gas, coal, and electricity, as well as CO_2 emissions from the use of fossil fuels, petroleum prices, energy reserves, and population; and data unit conversion tables.

(continued from page 37)

Developing new technologies that create energy by using fossil fuels more efficiently alone is not enough to control the emissions of greenhouse gases being emitted into the atmosphere. Without the increased use of *renewable* energy resources and the weaning from the dependence on fossil fuels, it will not be possible to bring global warming under control in time to prevent irreversible damage to the environment and life on the planet.

Emissions from Coal

According to "Stuck on Coal, Stuck for Words in a High-Tech World," a December 4, 2007, article in the *New York Times,* while society of the 21st century has progressed technologically in many ways over the past century, it is still stuck in the old-fashioned, outdated mode in popular energy choices. In particular, even with the new technology such as the growing forms of renewable energy, societies worldwide are still heavily dependent on coal and plan to remain that way despite the repeated pleas and warnings from climatologists about the life-threatening consequences of global warming if changes are not made now.

In the article Dr. James E. Hansen, head of the NASA Goddard Institute for Space Studies and a world-renowned climate expert, noted that the world's population, currently ignoring the issues of global warming, was "like society willingly ignoring a tragedy unfolding in plain sight in the same way millions of people did as the Holocaust swept Europe. The coal trains that serve higher CO_2 emitting power plants are like 'death trains.'"

Using coal-fired plants to generate electricity produces more greenhouse gases for each resulting watt than using oil or natural gas, but coal—often referred to as a dirty fuel—is attractive because it is relatively inexpensive. In countries where there are no emission controls (such as China and India), the coal industry today is booming. The International Energy Agency (IEA) projects that the demand for coal will grow by 2.2 percent a year until 2030, which is a rate faster than the demand for oil or natural gas. Yet, even with warnings to cut back on coal use and greenhouse gas emissions, according to the Union of Concerned Scientists (UCS) the United States currently has plans to increase its emissions by building many more coal-fired plants, with the majority of them lacking *carbon capture and storage* (CCS) technology—equipment that allows a plant to capture a certain amount of CO_2 before it is released and store it underground.

One of the most serious users of coal now and in the future is China. New figures from the Chinese government reveal that coal use has been climbing faster in China than anywhere else in the world. According to a report in the *Economist,* China opens a new coal-fired plant each week. Their rising energy consumption is making it more difficult to effectively slow the global warming process. The IEA in Paris predicts that the increase in greenhouse gas emissions from 2000 to 2030 in just China will be comparable to the increase from the entire industrialized world.

China is currently the world's largest consumer of coal, and its power plants are burning it faster than the trains can deliver it from the mines in China. As a result, it is also importing coal from Australia to meet its rising demands. It has also become the world's fastest-growing importer of oil. The Chinese are using more energy in their homes than ever before, and with a population of 1.25 billion people, energy use is expected to skyrocket. This increase in energy use will affect other energy-related sectors. China, for example, is now the world's largest market for television sets and one of the largest for other electrical appliances. It also has the world's fastest-growing automobile market. All of these commodities require the use of fossil fuels to manufacture and operate their industries, either directly or indirectly. Energy generated from coal and oil is expected to have a significant impact on global warming.

China is not the only country with a growing demand for energy. India, Brazil, and Indonesia are also showing a surging demand for energy. Power plants are burning increasing amounts of coal to meet the exploding new demands for electricity to serve both industry and private households. According to the New China News Agency, China's capacity to generate electricity from coal will be almost three times as high in 2020 as it was in 2000. China currently uses more coal than the United States, the European Union, and Japan combined. It should be noted that China is the world's leading builder of more efficient, less polluting coal power plants. With their increased expertise in building this technology, they have decreased the cost.

If China's carbon use keeps up with its current economic growth, their CO_2 emissions are projected to reach eight gigatons a year by 2030, an amount equal to the entire world's CO_2 production today. According to a report in the journal *Science,* China is completing up to two new coal plants per week in some areas, which are necessary to fuel their booming manufacturing expansion. For example, in 2000, steel production in China was reported at 140 million tons (127 million metric tons). In 2006, they produced 419 million tons (380 million metric tons). According to the International Iron and Steel Institute, in 2008 China's production headed the world at 489 million tons (444 million metric tonnes), which is twice as much as the United States and Japan combined. In addition to new construction, the steel is also being used for the manufacture of cars. In 1999, Chinese consumers bought 1.2 million cars. In 2006, 7.2 million cars were sold—an increase of 600 percent.

A report on June 11, 2006, in the *New York Times* stated that pollution from China's coal-fired plants is already affecting the world. In April 2006, a thick cloud of pollutants originating from northern China drifted on the air to Seoul, South Korea, then across the Pacific Ocean to the western United States. Scientists were able to track the progression and route of this brown cloud via real-time satellite imagery. According to researchers in California, Oregon, and Washington, a coating of sulfur compounds, carbon, and other by-products of coal combustion were found on mountaintop detectors in the Pacific Northwest.

Steven S. Cliff, an atmospheric scientist at the University of California at Davis, said, "The filters near Lake Tahoe in the mountains of eastern California are the darkest we've ever seen outside smoggy urban areas."

The sulfur dioxide produced during coal combustion poses an immediate threat to China's population, contributing to roughly 400,000 premature deaths each year. In addition, it causes acid rain that poisons rivers, lakes, wetland ecosystems, agricultural areas, and forest ecosystems. As the sulfur pollution is blown through the atmosphere globally, it causes global dimming elsewhere in the world. Also, the CO_2 coming from China will exist in the atmosphere for decades. According to the *New York Times*, "Coal is China's double-edged sword—the new economy's black gold and the fragile environment's dark cloud."

Based on a report from the Environment Maine Research & Policy Center, several domestic energy companies are planning on building more than 150 new coal-fired power plants in various locations across the United States. Environment Maine states that in addition to posing an environmental threat and increasing the environmental degradation from global warming, the addition of these coal plants are a threat to both energy security and a contributor to economic problems. Jennifer Andersen, an Environment Maine Research & Policy Center advocate, said, "We're lining up for a sprint in the wrong direction on U.S. energy policy. Expanding our dependence on coal will only worsen coal's impact on global warming emissions and intensify the other environmental impacts and economic risks from coal."

Based on information from the U.S. Department of Energy (DOE), the potential impacts of completing 150 proposed coal-fired plants would include the following:

- A 10 percent increase in U.S. global warming emissions.
- A 30 percent increase in U.S. coal demand; requiring the opening of new mines and expanding infrastructure for transportation.
- A $137 billion investment in dirty, outdated coal-burning technology. Only 16 percent of the proposed plants nation-wide would use coal gasification technology, and none would incorporate CCS. The majority would use old technologies that are already responsible for massive global warming emissions and the release of large quantities of pollutants responsible for human health problems.

"Companies that build coal-fired power plants today are gambling with their investors' money," said Leslie Lowe of the Interfaith Center on Corporate Responsibility, a coalition of investors promoting social responsibility. "They are betting that operating coal-fired power plants will continue to be cheap, despite the near certainty that global warm-ing pollution will be regulated within the lifetime of the plants."

These examples are not the only ones in the United States where coal is still being trumpeted as the primary choice for energy. In April 2006, the TXU Corporation announced its plans for eight new coal-fired plants to be constructed in Texas, making a total of 11 new coal-fired projects for that company—construction plans equaling 8,600 mega-watts in power and $10 billion in committed future capital investments. In June 2006, NRG Energy announced its intent to construct six new coal-fired projects to be built in Texas and Connecticut. In July 2006, the PacifiCorp electricity company announced plans for two new coal-fired facilities to serve their markets in Oregon. On *60 Minutes* on April 26, 2009, the CEO of Duke Energy, Jim Rogers, stated that although his stacks pump out 100 million tons (102 million metric tonnes) of CO_2 each year, he is planning two more plants this year.

Jennifer Andersen commented, "America could substantially reduce its global warming pollution using existing technology to improve energy efficiency and increase the use of clean, renewable energy sources such as wind, solar, geothermal, and biomass. What is more, these steps would be good for America's economy, creating jobs and improving productivity, but not if we stake our energy future on coal. Our leaders must take decisive action to stop the rush to build new coal plants and avoid the worst effects of global warming."

Emissions from Natural Gas

Of all the fossil fuels, natural gas is considered the cleanest. Composed primarily of methane, the principal products of its combustion are CO_2 and water vapor. During combustion, very small amounts of sulfur dioxide and nitrogen oxides are released, but virtually no ash or particulate matter is involved. The table below illustrates a comparison of emission levels of the various fossil fuels.

Because CO_2 makes up such a high proportion of greenhouse gas emissions, reducing CO_2 emissions would be significant in combating the greenhouse effect and global warming. The combustion of natural gas emits almost 30 percent less CO_2 than oil and just under 45 percent less CO_2 than coal.

One issue concerning the use of natural gas is whether the presence of methane—the principal component of natural gas and a very potent greenhouse gas—makes global warming significantly worse. According

Fossil Fuel Emission Levels (pounds per billion Btu of energy input)			
POLLUTANT	NATURAL GAS	OIL	COAL
Carbon Dioxide	117,000	164,000	208,000
Carbon Monoxide	40	33	208
Nitrogen Oxides	92	448	457
Sulfur Dioxide	1	1,122	2,591
Particulates	7	84	2,744
Mercury	0.000	0.007	0.016

Source: Energy Information Administration

to the EIA, although methane emissions account for only 1.1 percent of the total U.S. greenhouse gas emissions, they account for 8.5 percent of the greenhouse gas emissions based on GWP. Sources of methane emissions in the United States include waste management operations, agriculture, and the leaks and emissions from the oil and gas industry. In a major study conducted by the EPA and the Gas Research Institute (GRI) in 1997, done to determine whether the reduction in CO_2 emissions from increased natural gas use would be offset by a possible increased level of methane emissions, it was determined that the reduction in emissions from increased natural gas use strongly outweighed the detrimental effects of increased methane emissions. Therefore, the increased use of natural gas in place of other, dirtier fossil fuels can serve to lessen the emission of greenhouse gases.

Smog is another air quality issue. It is the primary constituent of ground level ozone and is formed by a chemical reaction of carbon monoxide, nitrogen oxides, volatile organic compounds (VOC), and heat from sunlight. The use of natural gas does not contribute significantly to smog formation, as it emits low levels of nitrogen oxides and virtually no VOCs. Because of this, it can be used to help combat smog formation in areas where ground level air quality is poor. The main sources of nitrogen oxides are electric utilities, motor vehicles, and industrial plants. Increased natural gas use in the electric generation sector, a shift to natural gas–operated vehicles, or increased industrial natural gas use have all been identified as ways to combat smog production, especially in dense urban centers where pollution is a serious problem. This is especially true in the summer, when natural gas demand is lowest and smog problems are the greatest. Industrial plants and electric generators could use natural gas in their operations in place of other more polluting fossil fuels, such as coal.

Particulate emissions—such as soot, ash, metals, and other airborne particles—also degrade air quality. A study conducted by the UCS in 1998 showed that the risk of premature death for residents in areas with high airborne particulate matter was 26 percent greater than for those in areas with low particulate levels. Natural gas emits virtually no particulates into the atmosphere. In fact, emissions of particulates from natural gas combustion are 90 percent lower than from the combustion of oil, and 99 percent lower than from burning coal. Therefore, the UCS determined that increased natural gas use in place of other dirtier hydrocarbons can help reduce particulate emissions in the United States.

Acid rain is another environmental problem that affects much of the eastern United States, damaging crops, forests, wildlife populations, and causing respiratory and other illnesses in humans. Acid rain is formed when sulfur dioxide and nitrogen oxides react with water vapor and other chemicals in the presence of sunlight to form various acidic compounds in the air. The principal source of acid rain–causing pollutants, sulfur dioxide and nitrogen oxides, are coal-fired power plants. Since natural gas emits virtually no sulfur dioxide and up to 80 percent less nitrogen oxides than the combustion of coal, increased use of natural gas would provide fewer acid rain–causing emissions.

As to industrial and electricity-generation emissions, which contribute greatly to environmental problems in the United States, the use of natural gas can significantly improve the emissions profiles for these two sectors. According to the National Environmental Trust (NET) in their 2002 publication entitled "Cleaning Up Air Pollution from America's Power Plants," U.S. power plants account for 67 percent of sulfur dioxide emissions, 40 percent of CO_2 emissions, 25 percent of nitrogen oxide emissions, and 34 percent of mercury emissions. Coal-fired power plants are the greatest contributors to these types of emissions.

Natural gas–fired electricity generation and natural gas–powered industrial applications offer a variety of environmental benefits, such as the following:

- Fewer emissions—natural gas emits lower levels of nitrous oxides, CO_2, particulate emissions, and virtually no sulfur dioxide.
- Reduced sludge—combustion of natural gas emits extremely low levels of sulfur dioxide, eliminating the need for scrubbers, and reducing the amounts of sludge associated with power plants and industrial processes.
- Reburning—by injecting natural gas into coal- or oil-fired boilers, nitrous oxide emissions can be reduced by 50 to 70

percent; sulfur dioxide emissions can be reduced by 20 to 25 percent.

- Cogeneration—the production and use of both heat and electricity can increase the energy efficiency of electric generation systems and industrial boilers.
- Fuel cells—natural gas fuel cell technologies are in development for the generation of electricity, where no emissions are involved.

The transportation sector is one of the greatest contributors to air pollution in the United States. According to the DOE, about half of all air pollution and more than 80 percent of air pollution in cities are produced by cars and trucks. Natural gas can be used in the transportation sector to cut down on these high levels of pollution from gasoline- and diesel-powered cars, trucks, and buses. According to the EPA, vehicles operating on compressed natural gas have reductions in carbon monoxide emissions of 90 to 97 percent and reductions in CO_2 emissions of 25 percent. Nitrogen oxide emissions can be reduced by 35 to 60 percent and other non-methane hydrocarbon emissions could be reduced by as much as 50 to 75 percent. Also, because the makeup of natural gas is relatively simple compared to traditional vehicle fuels, there are fewer toxic and carcinogenic emissions from natural gas vehicles and virtually no particulate emissions. These attributes argue forcibly for the use of natural gas in the transportation sector.

However, while it is true that burning natural gas emits fewer harmful pollutants and has fewer harmful effects than coal or oil, it must be acknowledged that this is only a short-term solution. The more effective long-term solution is the replacement of fossil fuels by renewable, clean energy sources.

Emissions from Oil

Ozone generated by exhaust emissions from vehicles is high in nitrous oxide and other particulates. Hot, humid days are especially dangerous to human health. Ground-level ozone created by exhaust can cause congestion, chest pains, and coughing. More serious health effects include severe difficulty in breathing for those who suffer from respiratory conditions such as bronchitis, emphysema, asthma, and heart disease. Health professionals claim that even healthy individuals can suffer when exposed consistently to a high ozone levels.

There are also serious environmental effects that result from the burning of oil, such as acid rain. When oil is burned, sulfur dioxides and nitrogen oxides are created in the atmosphere, and, when they combine with moisture, acid rain is created. When acid rain falls back to Earth, it enters rivers, lakes, and streams and can kill plants, fish, and wildlife. Buildings, statues, and other man-made structures can also be damaged as a result of the chemical reaction that occurs when acid hits them. Many historical buildings worldwide are being eroded by acid rain. As oil and gasoline are burned, CO_2 is also a by-product that enters the atmosphere, contributing to global warming.

HEALTH ISSUES ASSOCIATED WITH GLOBAL WARMING

One component of preparing for climate change involves planning for new or increasing threats to human health. As ecosystems change through intense weather events, shifting habitats, wildfires, heat waves, and other effects, humans must be prepared for inevitable changes. There is an overwhelming amount of evidence that rising levels of CO_2 in the Earth's atmosphere are having a serious impact on the climate with secondary effects on the Earth's physical systems and ecosystems. There have been increases in severe weather events, rising sea levels, migrations and extinctions of both plant and animal species, shifts in climate patterns, and melting of glaciers and *permafrost.* Now scientists are recognizing that global warming is the biggest threat to health of populations worldwide in the 21st century.

On May 13, 2009, a report from University College London was published in the *Lancet.* The report focused on the growing crisis in the environment and health care's role in it. Anthony Costello, a professor of international child health and director of the Institute for Global Health at University College London, said, "Climate change is a health issue affecting billions of people, not just an environmental issue about polar bears and deforestation. We are setting up a world for our children and grandchildren that may be extremely frightening and turbulent."

EMISSIONS AND OIL: QUICK FACTS

According to the Electric Drive Transportation Association (EDTA), the following facts about oil, emissions, and the environment were gathered from leading authorities worldwide:

- The United States' demand for oil is projected to increase by nearly 50 percent by 2025 (Hydrogen Posture Plan, U.S. DOE, February 2004).
- The Clean Commute program in New York, in which commuters drove *electric vehicles* to public charging outlets in rail stations, saved 6,664 gallons of fuel in one year (NK Clean Commute Program Report, U.S. DOE, August 2003).
- In 2000, the United States consumed almost 20 million barrels of oil products every day. Of that, 40 percent was used to fuel cars and trucks at a cost to consumers of $186 billion (Union of Concerned Scientists, January 2003).
- By 2020, oil consumption is expected to grow by nearly 40 percent and dependence on imports is projected to rise to more than 60 percent (Union of Concerned Scientists, January 2003).
- Transportation accounted for 66 percent of all oil consumed in the United States in 2000 (DOE, May 2001).
- Oil consumption is expected to rise to 25.8 million barrels per day (bpd) by 2020, mainly due to growth in consumption of transportation fuels (DOE, May 2001).
- A 10 percent reduction in energy use from cars and light trucks (achieved by introducing an alternative fuel or improv-

The repercussions will be felt first in the developing world, as can already be seen in African drought and South Asian flooding. However, climate change is on target to cause damage in the developed world, starting with the segments of the population at the greatest risk of health problems from the atmospheric changes associated with global warming, such as increased air particulates, greenhouse

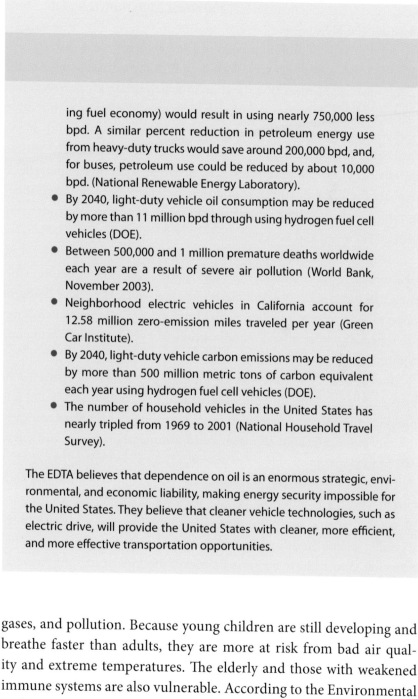

ing fuel economy) would result in using nearly 750,000 less bpd. A similar percent reduction in petroleum energy use from heavy-duty trucks would save around 200,000 bpd, and, for buses, petroleum use could be reduced by about 10,000 bpd. (National Renewable Energy Laboratory).

- By 2040, light-duty vehicle oil consumption may be reduced by more than 11 million bpd through using hydrogen fuel cell vehicles (DOE).
- Between 500,000 and 1 million premature deaths worldwide each year are a result of severe air pollution (World Bank, November 2003).
- Neighborhood electric vehicles in California account for 12.58 million zero-emission miles traveled per year (Green Car Institute).
- By 2040, light-duty vehicle carbon emissions may be reduced by more than 500 million metric tons of carbon equivalent each year using hydrogen fuel cell vehicles (DOE).
- The number of household vehicles in the United States has nearly tripled from 1969 to 2001 (National Household Travel Survey).

The EDTA believes that dependence on oil is an enormous strategic, environmental, and economic liability, making energy security impossible for the United States. They believe that cleaner vehicle technologies, such as electric drive, will provide the United States with cleaner, more efficient, and more effective transportation opportunities.

gases, and pollution. Because young children are still developing and breathe faster than adults, they are more at risk from bad air quality and extreme temperatures. The elderly and those with weakened immune systems are also vulnerable. According to the Environmental Defense Fund (EDF), there are four key health-related factors associated with global warming: heat waves, smog and soot pollution, food- and water-borne diseases, and stress from post-traumatic stress disorder (PTSD).

Infants and children four years old and younger are extremely sensitive to heat. When heat waves occur and children are subjected to the urban heat island effect (the situation in which urban areas are warmer than rural areas because asphalt pavements, buildings, and other human-made structures absorb incoming solar radiation and reemit the energy as longwave [heat] radiation), they face the risk of becoming rapidly dehydrated and can suffer the negative, sometimes fatal, effects of heat exhaustion or stroke. In addition, because young children's lungs are still developing, they can suffer irreversible lung damage from being exposed to smog and soot pollution.

As global warming increases in intensity and food- and waterborne diseases spread into areas where they have never existed before, children are much more susceptible to illness—especially those living in poverty. As extreme weather events leave families homeless, children are especially prone to the complications of PTSD as they try to cope with the upheaval of their lives.

By the year 2030, one-fifth of the U.S. population is projected to be older than 65. Because the elderly often have frailer health and less mobility, they are at greater risk from heat waves. If their incomes are limited and they cannot afford air-conditioning, their health and safety can be in jeopardy.

Another sector of the population at risk is composed of those with chronic health conditions. People with heart problems, respiratory illnesses, diabetes, or compromised immune systems are more likely to suffer serious health complications because of the effects of global warming. According to Dr. John Balmes of the American Lung Association of California, higher smog levels "may cause or exacerbate serious health problems, including damage to lung tissue, reduced lung function, asthma, emphysema, bronchitis, and increased hospitalizations for people with cardiac and respiratory illnesses." Smog forms when sunlight, heat, and relatively stagnant air meet up with nitrogen oxides and various volatile organic compounds.

Dr. John Balbus, chief health scientist at EDF, remarked, "The number of people with asthma in this country has more than doubled over the past 25 years, led by soaring rates in children. With climate change worsening smog in some areas and altering pollen levels, future air quality may pose a greater threat to our health, especially those of us with asthma and other lung diseases." Those with preexisting conditions such as weakened immune systems are also highly susceptible to catching diseases spread by mosquitoes or other vectors.

EDF has identified other groups as being potentially vulnerable to the negative effects of pollution, illness, and global warming. They include:

- pregnant women and their unborn children—may be unable to take medications or have access to air-conditioned locations
- people living in poverty—no access to air-conditioning or medical care
- people living in areas of chronic pollution—consistent exposure to unhealthy air compromises the system, leading to greater susceptibility to infectious diseases
- geographic areas where climate change has a great effect—storms, flooding, erosion, tornadoes, wildfires

According to a report issued by EDF, the table on the next page lists the U.S. regions most at risk from potential climate change.

Today, it is becoming increasingly common for cities to prepare emergency plans outlining actions that need to be taken if the negative effects of climate change and global warming affect their area. The National Oceanic and Atmospheric Administration's (NOAA) National Weather Service has put together a Heat/Health Watch Warning System for U.S. cities to improve forecasts and warnings for excessive heat. Many cities are already taking action. For further information, see www. noaanews.noaagov.com.

Topography also plays a role in an area's pollution. Cities located in close proximity to mountain ranges experience unique patterns of pollution. In areas where mountain ranges act as physical barriers and trap pollution over cities, air inversions are common, especially during the winter months. In valleys or on the lee sides of mountains, if a warmer air mass moves above cooler air, it traps the cooler, denser air underneath and increases the severity of air pollution. Los Angeles is an example of this, where the warm desert air from the east comes

Potential Regional Effects of Climate Change in the United States

GEOGRAPHIC REGION	POTENTIAL NEGATIVE EFFECTS
Southeast Atlantic and Gulf Coasts	Violent storms, strong storm surges, flooding, coastal erosion, damage to buildings and roads, contamination of drinking water
Southwest	Higher temperatures, less rainfall, arid climate, increased wildfires, worsened air quality
Northwest	Heavy rainfall, flooding, sewage overflow, increased spread of disease
Great Plains	Milder winters, scorching summers, decreased agricultural production, intense heat waves
Northeast	Higher temperatures, more allergies, spread of disease by insects and animals
Alaska	Melting permafrost, retreating sea ice, disturbed ecosystems, reduced subsistence hunting and fishing, milder temperatures, increase in insects and forest pests

Source: Environmental Defense Fund

over the mountains to the east of Los Angeles and lies over the cooler Pacific Ocean air. The cooler air becomes trapped because it cannot rise through the less dense warm air above it, and the pollution in the cold air accumulates. In mountain valleys, a similar situation occurs where warm air overlies the colder air that accumulates in the valleys. In cities, heat island effects are common. Warm air filled with pollutants collects and then spreads out over the nearby suburbs.

The greenhouse gases contributing the most to the anthropogenic greenhouse effect are listed in the following table.

Gases Contributing to the Anthropogenic Greenhouse Effect		
GAS	RATE OF INCREASE (% PER YEAR)	RELATIVE CONTRIBUTION (%)
CO_2 (carbon dioxide)	0.5	60
CH_4 (methane)	1	15
N_2O (nitrous oxide)	0.2	5
O_3 (ozone)	0.5	8
CFC-11 (trichloro-fluoromethane)	4	4
CFC-12 (trichloro-fluoromethane)	4	8

Mountains serve as effective barriers, allowing descending warm air to trap cold, polluted air close to the ground, making the air unhealthy to breathe. *(Nature's Images)*

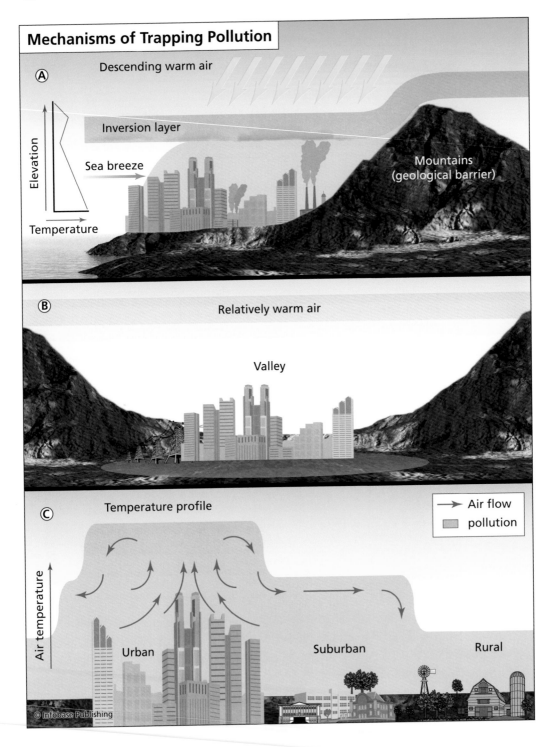

Mechanisms of Trapping Pollution

Based on research conducted at the Scripps Institution of Oceanography/University of California, San Diego, a new analysis of the pollution-filled brown clouds over southern Asia demonstrates that the region will be able to slow or stop some of the alarming retreat of glaciers only by reducing the existing air pollution.

Leading the research team, Veerabhadran Ramanathan, an atmospheric physicist at Scripps, concluded from the research, "The rapid melting of these glaciers, the third-largest ice mass on the planet, if it continues, will have unprecedented effects on southern and eastern Asia."

According to Achim Steiner, the executive director of the United Nations Environment Programme (UNEP) and United Nations undersecretary-general, "The main cause of climate change is the buildup of greenhouse gases from the burning of fossil fuels. But brown clouds, whose environmental and economic impacts are beginning to be unraveled by scientists, are complicating, and in some cases aggravating, their effects. The new findings should spur the international community to even greater action. For it is likely that in curbing greenhouse gases we can tackle the twin challenges of climate change and brown clouds and, in doing so, reap wider benefits from reduced air pollution to improved agricultural yields."

Jay Fein, program director in the National Science Foundation's (NSF) division of atmospheric sciences, remarked, "In order to understand the processes that can throw the climate out of balance, Ramanathan and colleagues, for the first time ever, used small and inexpensive unmanned aircraft and their miniaturized instruments as a creative means of simultaneously sampling clouds, aerosols, and radiative fluxes in polluted environments, from within and from all sides of the

(opposite page) This illustration depicts three different types of inversions: (a) a coastal inversion, where a sea breeze becomes trapped under descending warmer air; (b) a mountain-valley inversion, where a layer of warm air settles over colder air trapped in the valley; and (c) an urban heat island effect, where warmer air keeps pollutants contained above the urban area.

clouds. These measurements, combined with routine environmental observations and a state-of-the-science model, led to these remarkable results."

What the study was successfully able to reveal was that the effect of the brown cloud was necessary to explain temperature changes that have been observed in the region over the last 50 years. It also clarified that southern Asia's warming trend is more pronounced at higher altitudes than closer to sea level.

Ramanathan concluded, "The conventional thinking is that brown clouds have masked as much as 50 percent of the global warming by greenhouse gases through the so-called global dimming. While this is true globally, this study reveals that over southern and eastern Asia, the soot particles in the brown clouds are intensifying the atmospheric warming trend caused by greenhouse gases by as much as 50 percent."

According to work done by the World Resources Institute (WRI), it was not until the late 1940s when air pollution disasters occurred on two separate continents that public awareness began to grow concerning outdoor air quality and its effects on human health. Both the 1948 killer fog in Donora, Pennsylvania, that killed 50 people and the London fog of 1952, where roughly 4,000 people died, spurred investigations that determined that the widespread use of dirty fuels was to blame. This started governments' efforts to take the problem of urban air pollution seriously.

Since this time, many contaminants in the atmosphere have been identified as harmful, and serious efforts have been undertaken to clean up the atmosphere from harmful components. The most common and damaging pollutants include sulfur dioxide, suspended particulate matter, ground-level ozone, nitrogen dioxide, carbon monoxide, and lead. All of these pollutants are tied either directly or indirectly to the combustion of fossil fuels. Even though major efforts are under way today to clean polluted air over cities, many cities worldwide still lack a healthy air quality. An inventory completed by the European Environment Agency determined that 70 to 80 percent of 105 European cities surveyed exceeded World Health Organization (WHO) air quality standards for at least one pollutant. In the United States, an estimated

80 million people live in areas that do not meet U.S. air quality standards, which are comparable to WHO standards. Other areas that do not meet WHO standards include Beijing, Delhi, Jakarta, and Mexico City. In these cities, pollutant levels sometimes exceed WHO air quality standards by a factor of three or more. Some of the cities in China exceed WHO standards by a factor of six. Worldwide, WHO estimates that up to 1.4 billion urban residents breathe air exceeding the WHO guidelines and that the health consequences are considerable, with a mortality rate of 200,000 to 570,000 annually. In addition, the World Bank has estimated that exposure to particulate levels exceeding the WHO health standard accounts for roughly 2 to 5 percent of all deaths in urban areas in the developing world.

It is stressed, however, that these mortality estimates do not reflect the huge toll of illness and disability that exposure to air pollution brings on a global level. Health effects span a wide range of severity from coughing to bronchitis to heart disease and lung cancer. The most vulnerable groups are infants, elderly, and those suffering from chronic respiratory conditions, including asthma, bronchitis, or emphysema. In developing cities alone, air pollution is responsible for approximately 50 million cases per year of chronic coughing in children younger than 14 years of age. It has been postulated that the increased use of air-conditioning in buildings will exacerbate the incidences of Legionnaires' disease, sick building syndrome, and infectious airborne diseases, such as influenza and tuberculosis. Taking health effects into consideration when discussing global warming is critical because it affects the future of society.

Evolution since the Industrial Revolution

Civilization has progressed tremendously since the industrial revolution began in the mid-18th century. Until that time, society was largely agrarian, but as discoveries were made, inventions developed, and various industries started, people flocked to cities where the jobs were. This marked not only a turning point in human history, but in environmental history as well.

The anthropogenic influence on the Earth's natural greenhouse effect began in earnest at this time. People's lifestyles were forever changed with the invention and implementation of steam power, electricity, mechanization, development of fossil fuels, and implementation of industry as it is known today.

Factories began to be built to produce thousands of products, electricity had to be generated and transported to make the systems operate, and fossil fuels were used in order to transport goods from one place to another. Because of these demands on the environment, the past two centuries have seen a huge stress on the Earth's natural environment, causing global warming at a rate never before seen on Earth.

This chapter looks at this historic turning point, why and how it happened, key events in its history and evolution, as well as the effects of modernization. It also looks at the ramifications for today's environment and for future generations if the problems are not acted upon immediately. Finally, it explores what some are referring to as the green industrial revolution.

THE INDUSTRIAL REVOLUTION

Prior to industrialization, the standard of living was focused on substantive measures—the majority of people had to be concerned with producing their food for survival. In medieval Europe, for instance, roughly 80 percent of the labor force was employed in subsistence agriculture. When some countries were able to start importing goods, they experienced major changes in their economy, population distribution, vegetative cover, agricultural production, income, population levels, urban growth, distribution of the workforce, diet, and clothing.

The first transformation to an industrial economy from an agrarian one was called the industrial revolution. It did not occur at the same time everywhere. Industrialization through advancements in manufacturing processes first began in the northwest portions of England in the 18th century. It then spread to Europe and North America in the 19th century and to the rest of the world in the 20th.

As an energy source, the use of coal was symbolic of the beginning of the industrial revolution, and sadly coal has been one of the largest contributors to global warming. This period saw major changes in agriculture, manufacturing, production, and transportation. It not only affected the economic aspects of society, but the social aspects as well. The effects were far-reaching, influencing life in many ways.

It was at this time that manual labor was replaced with machine-based manufacturing. The beginnings of these advancements were seen in the textile industries, in the development of iron-making techniques, and in the increased use of refined coal. Simultaneously, methods of transportation were improved in order to expand the world markets.

Canals, roadways, and railways were all used to expand to new markets. Coal was used to create steam power and allowed production rates to climb drastically. Another major advancement was the development of all-metal machine tools. These made it possible to design additional machinery used in the manufacture of even more products. As the effects of this modernization spread around the world, the impact on societies was enormous.

What historians refer to as the first industrial revolution began in the 18th century. It then merged into the second industrial revolution around 1850. It was during this second wave that technological and economic progress gained momentum—mainly due to the development of assembly lines, steam-powered railways, and ships. Improvements were then made when the internal combustion engine and electrical power generation became available.

Throughout industrialization, some of the world's most important inventions were developed. Technological innovation was the key component of the industrial revolution, with its most critical invention being the Watt steam engine in the late 1700s.

It has been suggested that the industrial revolution started in Europe because Europe had easy access to resources such as coal near their manufacturing centers, as well as access to food and wood from the New World. In addition, investment capital was more accessible in Britain's economy at the time.

There are other inventions also associated with the industrial revolution. For example in the textile industry, cotton spinning was an important component. Three major spinning looms were invented, enabling the spinning of worsted yarn, flax linen, cotton, and other textiles.

The organization of labor also played a key role. This was when the assembly line work system was developed. By having a series of workers trained to do a single task on a product, then having it move along an assembly line to the next worker for their trained input, the number of finished goods was able to rise significantly, greatly improving efficiency and output.

The major change in the metal industries during this era was the replacement of wood with fossil fuel (principally coal) as a source of energy. It was at this time that wrought iron, steel, and the crucible

The Watt engine was one of the most famous inventions introduced during the industrial revolution. *(Tamorlan)*

steel method were developed. Mining advances were also made during the industrial revolution. The invention of the steam engine gave a tremendous boost to the development of mining. The steam engine enabled much easier removal of water from shafts, allowing mines to be dug deeper, letting more coal be extracted. In fact, it was the steam engine that greatly reduced the fuel costs of engines and made the mining industry much more profitable.

Other inventions associated with the era included Portland cement/concrete, machine tools, sewage systems, boring machines, milling machines, gas lighting, seed drills, threshing machines, and traction engines, to name but a few.

The industrial revolution marked a critical era when several ingenious inventions involved machines to harness energy to do work and be used in the production of various goods. At the time, the fact that burning coal could replace the need for human labor was an attractive concept. Unfortunately, little attention was paid to the environment. Therefore, while the industrial revolution meant that more goods could be produced for human consumption, it also marked a major turning point in the Earth's environmental health.

During this revolution and time that dramatically changed every aspect of human life in a seemingly positive way, the negative human impact on natural resources and energy use was also irreversibly put into motion. At the time, however, society was not focused on the negative impacts being inflicted on the environment, perhaps because the effects were not suddenly apparent. Beginning in the mid-1700s in Great Britain, when machinery began to replace manual labor, fossil fuels gradually replaced the traditional wind, water, and wood energy sources. It was not until about 200 years later that society began to realize the seriousness of the damage that had been done to the environment. As technology advanced, manufacturing became more efficient, new discoveries were made, and new products were developed. While this meant that more goods could be produced for human consumption, it also meant that more pollutants were continually being pumped into the atmosphere and more natural resources were being exploited to produce more goods.

One of the industrial revolution's most drastic effects was that as more inventions and commodities were developed, more jobs were created, and as more jobs were created, more people flocked to the cities in order to be employed at the factories located nearby. Along with this, world population growth exploded—world population by 1,000 C.E. was around 300 million. At the beginning of the industrial revolution, the population was roughly 700 million; by 1800, it was 1 billion. By 1850, the world population had expanded to 2 billion. During the 1900s, world population began growing exponentially to 6 billion people, a 400 percent increase. In terms of the environment, this population explosion contributed heavily toward rising air pollution.

Even worse than all these things was what the revolution did to the human mind-set. It changed the way people thought about themselves in relation to nature. Unfortunately, for many, it promoted the idea that humanity had finally mastered nature and was now apart from and above it. People began using energy resources—principally coal—without regard for any ill effects they might have on the environment. In fact, it was not until the 1960s that the issue came to public attention that perhaps this uncurtailed use of natural resources without regard for the effect on the environment was finally brought to public light and given the attention it deserved.

Rachel Carson, author of the 1962 book *Silent Spring,* brought people's attention to how their actions affected nature. She introduced the concept of sustainable production and development. Thus began the birth of the concept that although fossil fuels (mainly coal) were responsible for great advancements, they came at an extraordinary cost to the environment and ultimately to the health of all living things.

Factories poured black smoke into the air and waste products into water supplies, leading to the concept of dirty air. At the beginning of the industrial revolution, the carbon dioxide (CO_2) concentration in

Coal-fired plants belch pollution into the atmosphere from their smokestacks, contributing to global warming. *(UC San Diego)*

the atmosphere was approximately 280 parts per million (ppm). Today, it is approximately 387 ppm—at a level higher than at any time in the past 750,000 years—and is still rising, even though several areas have begun implementing measures to curtail carbon output.

In 1949, the American geophysicist M. King Hubbert predicted that the fossil fuel era would be very short-lived and that other energy sources would have to be found. He also predicted that oil would reach its peak production period in the 1970s and then enter a steady decline against the rising energy demands of a steadily growing population. Hubbert's theory—known as Hubbert peak—matches exactly what happened in the United States in 1971. Unfortunately, society's dependence on fossil fuels is so ingrained that now, when new energy sources must be found and lifestyles changed in order to use them, people are resistant to change.

As the scientific community looks back to the beginning of the industrial revolution, it is possible to see how emissions from human inventions have significantly changed the Earth's atmosphere, leading to global warming. While humanity's progress and inventions have benefited populations worldwide and their effects were largely unacknowledged or misunderstood at the time, the scientific community now understands just how tremendous an impact that technology over the past 250 years has had on the environment. Of all the sectors in the economy today, the transportation sector has been responsible for the greatest share of environmental damage. Not only have mining and drilling operations invaded and disturbed sensitive areas, such as the fragile polar ecosystems, but the worldwide damage from emissions has caused negative global effects.

In the mid-20th century, a few major events occurred that finally shocked people into becoming more aware of their environment and seeing a need to become environmentally responsible. In 1948, in the valley of Donora, Pennsylvania, pollutants from local coal plants combined with trapped air to produce a lethal cloud. In 1952, in London, sooty coal smoke and fog combined to produce killer smog, causing thousands of people to die. Around the same time, people began to notice that fish were dying and that the acidity levels of rain were high—leading to the discovery of acid rain. All three incidents were connected to changes occurring alongside global warming.

DONORA, PENNSYLVANIA, SMOG, 1948

The worst air pollution disaster in U.S. history occurred in 1948. Pollution from the U.S. Steel Corporation's Donora Zinc Works smelting operation and other sources containing sulfur, carbon monoxide, and heavy metal dusts was trapped by an inversion, where a warm air mass traps cold air near the ground. According to the Pennsylvania Bureau of Industrial Hygiene, the pollution came from a combination of the zinc smelting plant, the steel mills' open-hearth furnaces, a sulfuric acid plant, a slag dump, and several coal-burning steam locomotives and riverboats.

The industrial area of western Pennsylvania had created pollution problems for almost 100 years. By the mid-1900s, industrial pollution had blanketed the air. The valley of Donora, surrounded by hills, caused the dense pollution fog to remain close to the ground where people easily inhaled the dangerous chemicals. The air became so filled with pollution that people could not even see across the street.

Between October 26 and 31, 20 people were killed and more than 7,000 were hospitalized or became ill as a result of severe air pollution. This tragedy served to open people's eyes to how unhealthy environmental conditions can pay havoc with human health and set a chain of events in motion. Eventually, what happened at Donora helped prohibit the use of coal as a residential heating source, replacing it with clean natural gas. By 1952, diesel engines replaced coal-powered engines in locomotives and boats. By 1955, almost 97 percent of Pittsburgh's emissions were reduced, and the smog had cleared.

In addition, the Pennsylvania state government established the Division of Air Pollution Control in 1949 to study air quality and its effect on human health. Statewide clean air regulations were enacted in 1966 and in 1970. The Pennsylvania legislature passed an Environmental Bill of Rights, stating that, among other things, people had a right to clean air.

THE GREAT LONDON SMOG

Early in December 1952, a cold fog blanketed London. Because it was so cold, people began to burn more coal than usual in an effort to stay warm. A dense mass of cold air had formed in an inversion, trapping the air near the ground. As the coal was burned, the pollutants and particulates began to build up, but because of the inversion the air could not escape—it was held down at ground level forcing people to breathe it. The air became saturated with coal smoke. The type of coal being burned did not help matters either. The British were burning low-quality, high-sulfur coal for home heating because they were using their higher-quality coal as an export product to aid their ailing economy.

The smoke originated from several sources: London's industries, residential furnaces, and fireplaces. A deadly fog blanketed the Thames River Valley from December 5 to 9, 1952. The fog—actually smog—ended up being so thick that driving became nearly impossible because visibility was so poor. It also entered buildings when doors were opened, causing the cancellation of concerts, movies, and other events. It smelled distinctly of coal tar. At the time, however, because heavy fog incidents were already so common in London, no one paid much attention to it.

In the weeks that followed, however, the medical community began compiling statistics and discovered that the fog had killed about 4,000 people. The majority of the victims were those with existing respiratory problems, the very young, or the elderly. In the following weeks, another 8,000 people died. The death rate peaked at 900 per day on the eighth and ninth and remained above average until just before Christmas. The fog asphyxiated countless cattle in the region as well. Huge quantities of impurities were released into the atmosphere. On each day during the foggy period, 1,102 tons (1,000 metric tons) of smoke particles, 2,205 tons (2,000 metric tons) of CO_2, 154 tons (140 metric tons) of hydrochloric acid, and 15 tons (14 metric tons) of fluorine compounds were released into the atmosphere. Even worse, 408 tons (370 metric tons) of sulfur dioxide were converted into 882 tons (800 metric tons) of sulfuric acid. This incident played a critical role in awakening a greater awareness of the environment.

Interestingly, as long ago as the 13th century, air pollution was recognized as a public health problem in the cities and large towns of the British Isles, and the burning of coal was identified as the principal source. Four hundred years later, John Evelyn, author of *Fumifugium*, wrote in 1661 of the "hellish and dismal cloud of sea-coale" that lay

A man guiding a bus with a flaming torch through thick fog during the London smog of 1952 *(www.nickelinthemachine.com)*

over London and recommended that "all noisome trades be banished from the city." Unfortunately, no one paid any attention.

In 1952, the incident was so severe that it caught the world's attention and led scientists to take a serious look at the ramifications of all pollution and the very real, deadly effects it could have on people. As a result, new regulations were enforced restricting the use of dirty fuels in industry and banning black, sooty smoke. Three acts in particular came out of the incident: the Clean Air Act of 1956 and the Clean Air Act of 1968 and the City of London (Various Powers) Act of 1954.

ACID RAIN

Acid rain is a serious environmental problem that negatively affects lakes, streams, forests, plants, and animals within an ecosystem. Acid rain applies to a mixture of wet and dry deposited material from the atmo-

THE CLEAN AIR ACTS

Clean Air Acts refer to existing legislation whose goal is to reduce smog and air pollution. Their enforcement has helped various countries not only improve environmental and health conditions in their own countries, but worldwide as well.

The Clean Air Act of 1956 was put in place by the Parliament of the United Kingdom as a result of the Great Smog of London in 1952. This event caught the world's attention and drove the message home that pollution is a very real and deadly problem. In effect from 1955 to 1964, the act was administered under the ministry of housing and local government in England and the department of health in Scotland. The purpose of this act was to control pollution through the establishment of smokeless zones. This marked a beginning in the modern-day environmental movement where an effort was made to reduce the health risks and environmental degradation of industrial pollution.

Within the smokeless zones, only smokeless fuels were allowed to be burned. The result was twofold. In the smokeless zones, not only was smoke pollution reduced, but the use of cleaner coals and increased use of electricity and natural gas helped to reduce sulfur dioxide levels. In order to help the situation, power stations were relocated from urban to rural areas in order to distance any emissions from heavily populated areas. As a result, air pollution in cities was dramatically reduced.

The Clean Air Act of 1968 in England introduced the concept of requiring the use of tall chimneys for industries that burned coal, liquid, or gaseous fuels for energy. When this act was passed, it was already known that smoke

sphere that contains higher than normal amounts of nitric and sulfuric acids. It can originate from both natural and human-made sources. The emissions are usually sulfur dioxide and nitrogen oxides and can originate from volcanic eruptions, decaying vegetation, and fossil fuel combustion. In the United States, the biggest source of acid rain is from electric power generation that relies on the burning of fossil fuels, principally coal.

Acid rain occurs when the gases interact in the atmosphere with water, oxygen, and other chemicals to form various acidic compounds.

pollution could be controlled, but sulfur dioxide removal was another matter. The British government's philosophy with the act of 1968 was that the higher the chimney, the better the dispersal of the air pollution.

Canada and the United States have also implemented clean air acts. In Canada, their first Clean Air Act was put into effect in the 1970s in order to regulate asbestos, lead, mercury, and vinyl chloride air pollutants. That act was then superseded by the Canadian Environmental Protection Act in 2000. A second Clean Air Act was introduced in 2006, focusing specifically on smog and greenhouse gas emissions. The goal of this act was to reduce Canada's 2003 emission levels by 45 to 65 percent by 2050. The act provides for regulations on automobile gas mileage efficiency for 2011 and targets for ozone and smog levels for 2025. This act has met some opposition, however, with the point being argued that not enough is being done to fight global warming and the act needs to be tightened up to achieve more firm results at a faster pace.

In the United States, the Air Pollution Control Act was passed in 1955, followed by the Clean Air Act of 1963, the Air Quality Act of 1967, the Clean Air Act Extension of 1970, and Clean Air Act amendments in 1977 and 1990. Several individual state and local government agencies have enacted similar legislation, both implementing federal programs and filling in local gaps where necessary.

The Clean Air Act amendments of 1990 proposed emissions trading to deal with acid rain, ozone depletion, and toxic air pollution. It also set new auto gasoline formulation guidelines and requirements.

The resultant mixture is a mild solution of sulfuric acid and nitric acid. When sulfur dioxide and nitrogen oxides are emitted from power plants and other industrial sources, the prevailing winds can blow the acidic compounds long distances before they are deposited from the atmosphere to the Earth's surface.

Deposits can occur as either wet or dry. Wet deposition occurs if the chemicals enter humid, moist areas as acid rain, fog, or snow. Once the deposits reach the Earth's surface as acidic water, it flows over and through the ground and directly affects the plants and animals it comes in contact with. The ultimate damage it is able to do on the surfaces and targets it comes in contact with depends on the chemistry of the soil and the type of vegetation or other biomass the acidic water comes in contact with.

In dry areas, such as the southwestern United States, the acid chemicals may become incorporated into dust or smoke and settle to the ground where they stick to whatever they come to rest on—homes, cars, the ground, buildings, trees, or other vegetation. If the area then receives moisture through a rainstorm, the acid chemicals can be turned into solution and spread.

Acid rain can be highly destructive in the natural world. It can cause lakes and streams to become acidified, it can damage trees, and it can upset the chemical balance of sensitive soils. It also has negative effects on human-made objects. It can accelerate the decay of building materials and paints. Many historical statues built of marble are being continually damaged, such as the Capitol Building, the Jefferson Memorial, the Washington Monument, and the Lincoln Memorial in Washington, D.C. Structures in Europe are meeting the same fate. These structures are being slowly eaten away by the acid in the deposits. In addition to acid rain's destruction on the ground, it also reduces visibility in the atmosphere and can contribute to respiratory health problems.

The effects of acid rain are most evident in aquatic environments, such as lakes, streams, ponds, marshes, and wetlands. Besides receiving direct acidic deposition from rainfall, these areas are also collection locations for acidic runoff from the land. Most natural aquatic environments have a pH between 6 and 8 (7 is neutral). If more acid is introduced and neither the water nor soil can neutralize it, the water becomes more

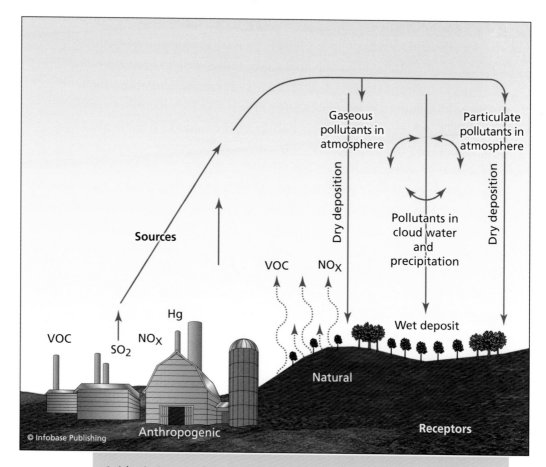

Acid rain is one of the environmental problems associated with pollution and global warming

acidic—in other words, it has a low pH level. If aquatic areas become too acidic, harmful effects are seen in fish. It can reduce their population numbers, completely eliminate fish species from a body of water, and decrease *biodiversity*. As acid rain flows through soils, aluminum is released from the soils into the lakes and streams. Both low pH and increased aluminum levels are directly toxic to fish. If these conditions do not kill the fish outright, they can lead to lower body weight, smaller size, and make the fish less able to compete for food and habitat.

The young of most species are more sensitive to environmental conditions than adults. At pH5, most fish eggs cannot hatch. At lower pH levels, some adult fish die. Some acid lakes have no fish in them. Even worse is the way it disrupts the *food chain*. To combat this problem, the decrease in sulfur dioxide emissions required by the U.S. Environmental Protection Agency's (EPA's) Acid Rain Program will significantly reduce acidification due to atmospheric sulfur. The EPA believes that without the reductions, the proportion of acidic aquatic ecosystems would remain high or dramatically worsen.

Over the years, foresters with the U.S. Forest Service have observed areas of slowed growth and trees with dry brittle leaves and needles. In some areas, entire trees have died off without an obvious reason. After much research, especially in the Appalachian Mountains, the Shenandoah Mountains, and the Great Smoky Mountains, it has been determined that the die-offs have been partly due to the negative effects of acid rain, pollution, and drought—all related to global warming.

Acid rain can be viewed as a problem that is often found in tandem with global warming. In order to stop it, it is necessary to clean up emissions from smokestacks and exhaust pipes and switch to alternative energy sources. Acid deposition is mainly caused by sulfur dioxide and nitrogen oxides when fossil fuels are burned.

Most of the sulfur dioxide emissions originate from coal-fired power plants, as does a large amount of nitrogen oxides. Approaches to clean up these emissions include using coal that contains less sulfur, washing the coal, and using scrubbers to chemically remove the sulfur dioxide from the gases leaving the smokestack. Power plants can also choose to discontinue the use of coal and use another energy source, such as natural gas, which has less sulfur dioxide. They may also use nonfossil fuel energy sources.

In order to cut nitrogen oxide emissions from automobiles, catalytic converters can be used. Emissions like these are monitored regularly by the EPA. They also make changes to the formulation of gasoline in order to allow it to burn cleaner. Alternative energy sources can be used to produce electricity, such as nuclear power, hydropower, wind energy, solar energy, and geothermal energy. All sources of energy have environmental costs as well as benefits, and some are still experimental at this point or simply still too costly. But as technology improves and advancements are made, it will be necessary to make energy choices that do not contribute to global warming.

COAL PLANT POLLUTION

In a study on coal plant pollution jointly commissioned by the Pew Center, Pace University, the National Environmental Trust, and the U.S. PIRG Education Fund, it was determined that while the entire population's health is at risk from exposure to fine particles, it is by far young children, the elderly, and those with respiratory disease who are at the greatest health risk.

Children face a certain risk because they breathe about 50 percent more air per pound of body weight than adults do. In addition, their respiratory systems are not fully developed, which makes them more susceptible to pollution, poor air quality, and other environmental threats. If children are exposed to particulates from coal mining and other pollutants, it can halt their lung development. Exposure to fine particles is associated with increased frequency of childhood illnesses.

Infants are especially susceptible to the fine particulates associated with coal plant pollution. In fact, a recent study found a 26 percent increased risk for sudden infant death syndrome (SIDS) in cities with high levels of particulate pollution. In addition, infants in high pollution areas were 40 percent more likely to die of respiratory causes. Children who suffer from asthma are also bothered by breathing fine particulate matter in the air.

Most of the coal used in the United States today is burned by power plants for the production of electricity. The oldest coal-fired facilities produce the largest share of the particle-related air pollution. Approximately half of all the power plant boilers in the United States are fueled by coal, and coal-burning power plants account for almost 90 percent of the sulfur dioxide emitted by all power plants.

In the study, several coal-burning power plants were tested for their emissions and particulate matter. The majority of power plant emissions were dominated by sulfate-acidic fine particles. The Pew Center concluded that in the United States, power plants exceeded all other polluters as the largest source of sulfate air pollution in the country and that they were responsible for two-thirds of the annual sulfur dioxide and one-fourth of the nitrogen oxides emitted in the United States each year. Furthermore, when released, these gas emissions form fine particles as they chemically convert in the atmosphere to form sulfate and nitrate particles. In addition, power plants also emit fine carbon soot particles directly from their smokestacks.

Even more disturbing is what the study revealed about long-term trends despite the implementation of the Clean Air Acts. According to the report, while the 1990 Clean Air Act Acid Rain Program resulted in significant improvements in atmospheric quality at first in reducing sulfur dioxide emissions from power plants, the emissions began slowly rising in years afterward. The National Emissions Trends Report subsequently documented that power plant sulfur dioxide emissions had crept upward every year since 1995, rising more than 10 percent. In fact, in 1998, power plants in the United States emitted 1.26 million more tons (1.14 million metric tons) of sulfur dioxide than they emitted in 1995.

In addition, nitrogen oxides particulate matter and volatile organic emissions have all crept up slightly over the past decade despite efforts to curb their emission levels. According to the study, "While the increases are not enormous, the data suggest poor progress in curbing power plant emissions."

The majority of the fine particle pollution does not originate from direct emissions, but from the conversion of sulfur dioxide and nitrogen oxides into fine particle sulfate and nitrate. The impact of these emissions is greatest in the midwestern United States (because it is densely populated with coal-burning power plants) and the eastern United States (because it is directly downwind of the majority of the nation's coal-burning power plants).

According to the study, because of the association between human health, presence of fine particles in the atmosphere, and coal-fired electric generation, health researchers have estimated the relative contribution of power plants to human mortality rates. Researchers at the Harvard School of Public Health have estimated that power plants are responsible for approximately 15,000 deaths per year. In addition, in the EPA's cost-benefit analysis of the Clean Air Act, they determined that, "The health benefits associated with reductions in power plant–generated fine particle pollution provided strong justification for pollution control costs imposed by the Act." To support their claims, a study conducted by the Harvard School of Public Health of two coal-fired power plants in Massachusetts determined that fine particle pollution from the plants was most likely associated with more than 100 deaths annually.

MODERNIZATION AND GLOBAL WARMING

When the developing nations began industrializing 250 years ago, thoughts of pollution, atmospheric degradation, and environmental damage did not enter the collective consciousness. Their newly gained conveniences, freedoms, and comforts were so attractive and welcome that very few people stopped to consider that what was actually going on was the beginning of an uncontrolled experiment. At the time, an answer to "how much CO_2 and other gases could humans pump into the atmosphere before there would potentially be too steep a price to pay?" was not a concern to the general populace.

Fortunately, not all were silent. One scientist did step forward. In 1896, the Swedish chemist Svante Arrhenius warned the world that CO_2 in time could heat up the Earth's atmosphere. Even then, however, it did not spark enough serious concern to trigger responsible environmental action.

Today, even though global warming has caused an increase in temperature of only about 1°F (0.6°C) over the past 100 years, that small increase has caused dramatic changes around the world: Ice is melting at unprecedented rates at both the North and South Poles, glaciers are rapidly melting, frozen tundra is thawing, sea levels are rising, animal species are migrating in search of temperature stability, the timing of seasons is changing, and 17 of the 18 hottest years on record have occurred since 1980.

The current environmental movement is gaining momentum, and the general public is becoming more aware of their impact on the environment since the beginning of the industrial revolution. This is due to the research efforts of the more than 2,500 scientists associated with the Intergovernmental Panel on Climate Change (IPCC) and dozens of environmental organizations such as the World Wildlife Fund, the Environmental Defense Fund, Greenpeace, and many others.

ENVIRONMENTAL TIME LINE

The following events that occurred during the period of the industrial revolution have had a significant effect on climate change.

1760   First experiments on use of coal gas for lighting by coal mine owner George Dixon in Newcastle, England.

1770   William Wordsworth, first of the English romantic poets, is born. Wordsworth thought the industrial revolution was an "outrage done to nature" and was appalled that the common people were no longer "breathing fresh air" or "treading the green Earth."

1775   The English scientist Percival Pott finds that coal is causing an unusually high incidence of cancer among chimney sweeps.

1784   Benjamin Franklin notes that the switch from wood to coal had saved what remained of England's forests, and he urged France and Germany to do the same.

1792   William Murdock first uses coal gas to light a small room in Redruth, Cornwall. He improved the gas by passing it through water—marking the beginning of the manufactured gas industry, which created vast pools of toxic coal tar in thousands of European and American towns and cities.

1799   Philippe le Bon becomes the first to illuminate a public building with gas.

1804   Smoke in Pittsburgh—complaints about coal smoke from many buildings such as blacksmith shops are filed. The smoke affected the "comfort, health, and . . . peace and harmony of the new city." As in other cities, polluters just built higher chimneys.

1812   First gas lights introduced in London by the Gas Light & Coke Co. This "town gas" would be used in every major U.S. and European city, but residential coal tar would remain an environmental problem into the 21st century.

1819   British parliamentary committee expresses concern that steam engines and furnaces "could work in a manner less prejudicial to public health."

1819   John Ruskin, an Oxford professor, detested the industrial revolution. He said modern towns were, "little more than laboratories for the distillation into heaven of venomous smokes and smells, mixed with effluvia from decaying animal matter, and infectious miasmata from purulent disease . . . [Every river was] a common sewer, so that you cannot so much as baptize an English baby but with filth, unless you hold its face out in the rain, and even that falls dirty."

1853 The novelist Charles Dickens opens *Bleak House* with an image of London as a twisted, twilight world of smoke, shadows, and wraiths. Dickens writes: "Smoke lowering down from chimney-pots, making a soft black drizzle, with flakes of soot in it as big as full-grown snow flakes—gone into mourning, one might imagine, for the death of the sun."

1859 Svante August Arrhenius was born February 19 in Sweden. The Nobel Prize–winning chemist was the first to predict global warming from fossil fuel–induced CO_2 buildup.

1863 Air pollution from the British chemical industry spurs the Alkali Act, intended to create reductions in hydrogen chloride emissions during alkali production. It allows agents of the first British pollution control agency, the Alkali inspectorate, to question industry officials and suggest improvements, but there were no actual regulations concerning amounts of air pollution until the act was revised in 1906.

1870 First coal mine safety laws passed in Pennsylvania following a fire that suffocated 179 men.

1873 December—first of a series of killer fogs in London. More than 1,150 die in three days. Similar incidents in 1880, 1882, 1891, 1892, and later.

1880 January inversion leads to another killer fog in London with 700 deaths.

1880s First U.S. municipal smoke abatement laws aimed at reducing air pollution from factories, railroads, and ships. Regulation under local boards of health under common-law nuisance statutes.

1881 Norway tracks first signs of acid rain on its western coast.

The negative effects of fossil fuels (coal, oil, and gas) can be seen as early as the mid-1700s with the beginning of the industrial revolution. These effects have continued ever since, compounding and adding to the problem of global warming.

Source: Dr. Bill Kovarik, Radford University, Virginia

Finally, in 2009, people are starting to focus on environmental issues. With the new Obama administration in place, slow change is starting to happen. On May 19, 2009, the administration proposed tough standards for tailpipe emissions from new automobiles, the first nationwide regulation for greenhouse gases. And the timing is just right—because the rate of temperature increase is accelerating.

The IPCC reported in their 2001 assessment that the Earth's temperature could climb by as much as 7°F (4.2°C) by 2050 and by as much as 10°F (6°C) by the year 2100—equaling a rate more than 60 percent

Countries Now Becoming Industrialized

© Infobase Publishing

These are countries that are now becoming industrialized. China and India are distinguished as darker red because they also represent major emerging economic countries. The gray countries are the rest of the world, either already developed or remaining undeveloped. If these newly industrializing countries rely on fossil fuels for the generation of energy, the impact to global warming will be devastating.

higher than they had predicted six years earlier. Some criticize this prediction, claiming it may actually be too low.

The problem facing society today is that drastic changes must be made—CO_2 emissions must be reduced. The effort must be coordinated worldwide; the efforts of only a few countries will not solve the problem. Despite societies' best attempts at controlling present and future emissions, however, global warming is not a problem that will just go away. Because of the long life span of several of the greenhouse gases, there will be effects left in the atmosphere for a long time to come—hundreds of years and more. The reality is that the industries of the world cannot convert from fossil fuels to clean fuels overnight. In addition, even with attempts by industries to reduce CO_2 emissions (which those in the United States have largely resisted), levels of CO_2 emissions are still continually rising. Since the industrial revolution, CO_2 levels have risen from 280 to 380 ppm. It will take a unified worldwide effort to curb CO_2 emissions—which may very well prove to be one of the most difficult challenges mankind has ever had to face.

THE GREEN INDUSTRIAL REVOLUTION

As global temperatures continue to climb and scientists worldwide warn of an impending climate crisis, the world also faces a financial crisis. According to "Time for a Green Industrial Revolution," an article by Nicholas Stern in *NewScientist,* now is an ideal opportunity to take advantage of the current technological revolution that has the ability to support the sustainable growth and development of a low-carbon global economy.

The switch to a green economy is especially important as more information is released that the risks and potential costs of continuing business as usual are even greater than originally thought. CO_2 emissions are growing more rapidly than initially projected, natural absorption of CO_2 by the Earth's ecosystems is lower than what was originally projected, and the physical impacts—such as glacier melt—are manifesting themselves much faster than originally predicted.

In the past, global climate research indicated that greenhouse gases needed to be stabilized at 550 ppm in order to keep the risks of potentially catastrophic impacts from occurring. According to the *Stern Review on the Economics of Climate Change,* however, it was recommended that CO_2 levels not be allowed to exceed 500 ppm if there was to be any hope of not permanently damaging the Earth's fragile ecosystems. The *Stern Review on the Economics of Climate Change* is a 700-page report that was released on October 30, 2006, by the economist Lord Stern of Brentford for the British government. It discusses the effects of climate change and global warming on the world economy. According to the *Stern Review,* annual global emissions must peak within the next 15 years before falling to half their 1990 level by 2050. Beyond that, it will be necessary to limit human additions to atmospheric greenhouse gases to under 10 gigatons of CO_2-equivalent per year, compared with the current 45 gigatons.

Although these reductions present a significant challenge and are criticized by many opponents to green energy, researchers of renewable green energy say new green technology is now affordable and manageable and that if 2 percent of the global gross domestic product (GDP) were committed each year for its development, it would not only be feasible, but a better investment in the long run to make the switch now to green technology.

In December 2009, the industrialized nations of the world are scheduled to meet in Copenhagen, Denmark, at the annual United Nations Framework Convention on Climate Change (UNFCCC). If progress can be made in Copenhagen for agreements to cut emissions by 80 percent by 2050 compared to 1990 levels, it would be a pivotal achievement in successfully combating global warming before it is too late.

Nicholas Stern is the British academic and economist who chairs the Grantham Institute for Climate Change and the Environment at the London School of Economics. He states that with the technological revolution moving at its current pace with the advent of computer technology, advancement of electronics, and the information revolution, scientific research and higher education are now seen as the new engine of current and future modernization. Grantham brings together international expertise on economics, finance, geography, the environment, international development, and political economy and has established a leading center for research and training in climate change and the environment. With present-day innovations advancing at a rapid pace, the time is prime for the development and dissemination of green technology or low-carbon technologies. Nicholas Stern is also the author of the *Stern Review on the Economics of Climate Change.* Recruited by Gordon Brown, then Chancellor of the Exchequer, he was appointed to conduct reviews on the economics of climate change and development. The *Stern Review* gained worldwide attention for its conclusions that addressing global warming now is the best economic choice. This report and his current role have placed him at the front of the pack of people aiming to solve the problems of global warming.

In order for the new green industrial revolution to gain momentum, the political arena must also evolve. Governments will need to be supportive. Policies and measures that remove barriers and provide incentives for developments in technology must be set in place.

Stern's article in *NewScientist* made three recommendations to promote green industries. The first is to spread existing low-carbon technologies, such as green household appliances. Second, it is important to provide for the development and upscaling of technologies that could become commercially viable within the next 15 years. This includes the need for improving and using carbon capture and storage (CCS). CCS is seen as a critical option for newly developed countries such as China and India, which currently rely on coal-fired power plants. The third recommendation is to encourage new breakthrough technologies that will lead to cuts in major emissions beyond 2030. In order to achieve this, energy research and development efforts should be doubled now.

Stern believes that the current downturn in the global economy represents a good opportunity to invest in low-carbon technologies because costs would be lower. Building these technologies would provide new job opportunities in the construction sector. Another favorable point is that if investments for systems that improve energy efficiency are made now and put in place, they will yield benefits in the future as utility costs for heating and cooling fluctuate through highs and lows. In the long term, investments in low-carbon technologies could provide for a sustainable and well-founded economic growth. According to Stern, "Continued unchecked emissions and high-carbon growth are not sustainable. In 2009, we have a real chance to set a path towards a low carbon future. It is the only realistic future for growth and for overcoming world poverty and the policies of governments worldwide. The next few months will set the tone for whether or not the opportunity offered by the recession can be harnessed. If the private sector does not get clear signals about where policy is going, it will be a major setback. If you lose the moment, people may feel that 'well, we'll come back to this.' That is not an option. Delays mean more carbon dioxide accumulating in the atmosphere, and a greater problem to address further down the road."

Global Warming and Pollution: Buildings and Homes

Global warming–associated pollution comes from three primary sources: electricity, heating, and transportation. The U.S. Environmental Protection Agency (EPA) estimates that more than 120 million Americans breathe unhealthy, polluted air each year. In order for the United States to make the necessary reductions in emissions that scientists are advising are necessary to curb global warming before permanent, unstoppable damage is done—cutting emissions by 15 to 20 percent by 2020 and by 80 percent by 2050—it will require serious changes in many areas of America's economy. Necessary changes include the increased use of clean, renewable energy and dramatic improvements in the efficiency with which energy is used in businesses and homes. The good news is that technology exists today that can help solve some of these challenges, and researchers are currently busy looking for even better solutions. Many cities in the United States, as well as countries around the world, are taking the issue of global warming seriously and already making the necessary changes to reduce global warming–associated pollution. By taking positive action, several benefits are being realized—a decreased dependence on fossil fuels, cleaner air, healthier communities, economic growth, and new jobs.

This chapter discusses energy use and efficiency in buildings and homes. It then explores ways everyone can make a difference in the environment where they live. Finally, it presents reasonable solutions to solve this growing problem before it is too late.

GREEN BUILDINGS

In the past, traditional buildings have not been designed or constructed with conservation in mind. Many have had a negative impact not only on the environment, but also on their occupants. Today, as global warming becomes better understood and society in general becomes more environmentally aware, a greater percentage of the public is looking at the concept of sustainable green energy projects— either as renovations of existing structures or as newly constructed structures.

The concept of a *green building* is one that is designed to preserve the health of those who occupy it, to conserve natural resources, and to reduce both energy and transportation costs. These objectives are accomplished by

- using sustainable building materials
- using materials (such as carpets) that emit fewer health-endangering toxins
- using water- and energy-conserving fixtures and appliances
- locating sites in proximity to services and public transportation
- making site improvements that reduce storm water runoff
- using energy-conserving, resource-responsible materials

Nongreen traditional buildings were not only designed without the environment in mind, but they are also generally expensive to operate, contributed to excessive resource consumption, generated excess waste, and added to pollution levels. Many cities nationwide and worldwide are now addressing the issues of excess energy consumption, resource conservation, and environmental sustainability in an effort to combat global warming.

The United States is home to about 5 percent of the world's population, yet it consumes approximately 25 percent of the world's energy and generates high levels of global warming–associated pollution. Urban sprawl has also led to increased traffic congestion and increased pollution.

INCREASING PRODUCTIVITY—REAL-LIFE EXAMPLES

Several businesses that have remodeled to become green have determined that it has had a positive effect in increased worker productivity and better economic returns. According to the U.S. Department of Energy (DOE) and the Rocky Mountain Institute, a study of the effect of office design on productivity found a direct correlation between specific changes in the physical environment and worker productivity. Some of the specific cases are summarized below.

Boeing—the manufacturer of aircraft—participates in the EPA's voluntary Green Lights program to promote energy-efficient lighting. Currently, they have retrofitted 13 percent of their 8-million-square-foot (743,224 m²) facility, reducing electricity use by up to 90 percent in some of its plants. They calculate their overall return on investments in the new lighting to be 53 percent—the energy savings paid for the lights in just two years. Lawrence Friedman, Boeing's conservation manager, believes that "if every company adopted the lighting Boeing has installed, it would reduce air pollution as much as if one-third of the cars on the road today never left the garage."

Boeing has also discovered some other interesting results. With their more efficient lighting, the employees have noticed that the glare inside the work area has been reduced. Friedman said, "The things the employees tell us are almost mind-boggling. One woman who puts rivets in 30-foot (9 m) wing supports had been relying on touch with one part because she was unable to see inside. Now, for the first time in 12 years, she could actually see inside the part." Friedman also said, "Most of the errors in the aircraft interiors that used to slip through weren't being picked up until installation in the airplane, where it is much more expensive to fix. Even worse, some imperfections were found during the customer walk-through, which is embarrassing and costly. Although it is difficult to calculate the savings from catching errors early, a manager estimated that they exceeded the energy savings for that building."

At the Pennsylvania Power and Light Company, their older lighting system was causing indirect glare from the work surfaces into employees' eyes, making their work less efficient—it took them longer to complete tasks and caused an increase in the number of errors they made. Russell Allen, superintendent of the office complex, said, "Low-quality seeing conditions were also causing morale problems among employees. In addition to the [glare and lighting] reflections, workers were experiencing eye strain and headaches that resulted in sick leave."

The power company decided to invest in modern, energy-efficient, nonglare track lighting that gave control to each individual workstation so that each employee could adjust their own lighting to their comfort and efficiency level. The results were noticed immediately. According to Allen, "As lighting quality is improved, lighting quantity can often be reduced, resulting in more task visibility and less energy consumption."

When Allen did a cost analysis on the effects of the lighting change he was surprised. He reported that the lighting energy use dropped by 69 percent and total operating costs fell 73 percent. The annual savings alone from the reductions in energy operating costs completely paid for the lighting system in less than four years—a 25 percent return on their investment. In addition, the newer lighting lowered heat loads (because of better efficiency), resulting in lower space cooling costs. Employee productivity also rose 13.2 percent. The savings in salary due to the increase in employee productivity paid for the new lighting in just 69 days. According to Allen, "Not only is this an amazing benefit, it is only one of several."

Before the upgrade, employees used an average of 72 hours of sick leave a year. Because the new lighting relieved eye fatigue and headaches, as well as boosted morale, the absenteeism dropped 25 percent. The improved lighting also reduced the number of errors employees made, producing overall higher-quality products. Allen concluded, "Personally, I would have no qualms in indicating that the value of reduced errors is at least $50,000 a year. If this estimate were included in the calculation, the return on investment would exceed 1,000 percent."

A new Wal-Mart in Lawrence, Kansas, opened in 1993. The building was a new prototype they called the eco-mart,—an experimental new design. Included in the concept was the use of native species for landscaping, a constructed wetlands for site runoff, as well as a source of irrigation, a building shell design for reuse as a multifamily housing complex, a structural roof system constructed from sustainable harvested timber, an environmental education center, and a recycling center. A major goal of the project was to design for energy efficiency. The building has a glass arch at the entrance for daylighting, an energy-efficient lighting system, an HVAC system that uses ice storage, and special light-monitoring skylights developed specifically for the project.

The initial cost for the construction of the eco-mart was about 20 percent higher than the cost for the standard Wal-Mart structure for several reasons: The roof was 10 percent more expensive because sustainable harvested timber was used; the cooling system was more expensive than usual; the building included a recycling center that other Wal-Marts did not have; and light-monitoring skylights were added that other Wal-Marts lacked. Taking these extra costs into account, Wal-Mart decided to install the skylights on only half the roof, leaving the other half without daylighting.

They received some interesting results when everything was completed and the store was in full operation. According to Tom Seay, Wal-Mart's vice president for real estate, "The sales rates in the portion of the store was significantly higher for those departments located in the day-lit half of the building. Sales were also higher than for the same departments in other stores. Additionally, employees in the half without the skylights are arguing that their departments should be moved to the day-lit side. Wal-Mart is now considering implementing many of the eco-mart measures in both new construction and existing stores."

ENERGY EFFICIENCY

According to the Union of Concerned Scientists (UCS), buildings account for about one-third of the energy consumed in the United States. Heating and cooling systems consume about 60 percent of the total, and lighting and appliances use the other 40 percent. On top of this, additional energy is required to manufacture and transport the building materials.

When buildings are designed, if they are designed with conservation of resources and energy in mind, they can be constructed to have less of an impact on global warming. For example, by using natural breezes and the Sun's energy and light and by using solar water-heating systems, the energy use in buildings can be reduced significantly. Not only do these options save money, but they help the environment and strengthen the economy by reducing the need for fossil fuels.

According to research done by the UCS, new buildings constructed in the United States today are more energy efficient than they have ever been in the past, as society has become more energy conscious. In order to reduce energy, several options can be employed. Sunlight, landscaping, natural breezes, and the choice of building materials can all reduce energy needs. One concept called passive solar design—the use of a building's structure to capture sunlight and store heat—can save up to 50 percent or more of the energy used in a building.

Another energy-saving practice concerns the physical orientation of a building. If a structure is oriented with its long face within at least 30 degrees of true south, it saves energy. Placing a high percentage of the windows on the south side also saves energy. Conventional buildings generally have about one-quarter of the windows on the south side, which averages about 3 percent of the building's total floor area. Optimally, if the total area of south-facing windows is at 7 percent, it allows the building to use more of the Sun's energy by absorbing it into the materials of the building. This, in turn, can save up to 25 percent of the building's conventional heating and, with protection from a shading overhang, can help reduce summer cooling expenses. During the winter, when the Sun is positioned at a lower angle on the horizon, the south-facing glass lets in sunshine to heat the space. During the summer, when the Sun is high in the sky, an overhang can prevent unwanted heat gain.

There are other low-cost passive design features that save energy as well. Ideally, the interior of a building should allow for the natural flow of heat in the winter and for the enhancement of ventilation during the summer. Using the south portion of the building for principal occupancy and the north and west for storage also helps conserve heating and cooling expenses.

The windows on the three nonsolar sides (north, east, and west) of a building also need special consideration. North-facing windows offer even lighting, but they also lose the most heat in the winter. The east- and west-facing windows, because of the low morning and evening Sun, can produce the highest air-conditioning demands. West-facing windows transmit large amounts of heat during summer afternoons and can cause excessive overheating. Windows that are double-paned, high-efficiency rated, and have a low-emittance- (Low-E) coated glazing should be used.

In order to enhance passive cooling, a building's windows should also be designed and placed to capture the prevailing winds. A casement window that opens to the wind can allow breezes into the building; a window placed on the opposite side of the building allows the stagnant air caught inside to flow out. In addition, the use of breezes for cooling can be used along with hot air's natural tendency to rise. If a high point is designed in a building, the natural movement of hot air to the ceiling can be combined with an opening to the outside, which can act as a release vent during the summer. The same principle can be used in the evenings to bring cool air into the building at night. The color of building materials is also important. In hot climates, the roofs should be light colored so that they reflect the Sun's direct heat instead of absorbing it into the building.

Rooftops are currently important sites for electric energy resources of the future. In several parts of the country, electric utilities are currently analyzing the possibility of renting rooftops for the placement of photovoltaic (PV) cells. These cells convert sunlight directly into electricity, and the system can be tied in directly to the utility grid. This means that when the PV system produces more electricity than the building is using, that electricity flows back into the utility's wires and is purchased by the utility company at a fair, competitive rate. When the building requires more electricity than it is currently producing through the PV, it is purchased from the utility in conventional fashion.

Hawaii's Mauna Lani Bay Hotel has acres of roof space. Working with PowerLight Corporation, they installed a PowerGuard® system of insulating photovoltaic (PV) roofing tiles that covers 10,000 square feet (929 m²) and generates 75 net kilowatts of electricity. This system will save the hotel enough in utility bills to pay for itself in five years. This building is an example of building-integrated photovoltaics (BIPV). *(SunPower, DOE/NREL)*

There are also active systems put in buildings today to create better energy efficiency. Active systems generally involve moving parts that circulate air through a building or move a liquid—most often water. A fan is a common method to pull hot air to other locations in the building. Warm air is often brought to a low point in the building in order to create an assisted convective loop, helping to circulate air throughout the building. Roofs can also support active systems, such as solar water heaters. The possible future use of the roof for PV and solar water heaters is another reason why roofs should face south.

Daylighting (the use of natural light in a building) reduces the need for electric lights and improves the visual qualities of an area. This is also consistent with the heating and cooling aspects of passive design. It is also possible for carefully controlled daylight to provide all of the necessary interior lighting with less heat emitted to the interior spaces that is released by incandescent or fluorescent lights. Because electrical lights create an excess amount of internal heat, they can cause air conditioners to be on through much of the year. Open interior plans that enable natural light to penetrate to all parts of the structure are important in commercial buildings, where electric lights are in the highest demand.

Using recycled office supplies also helps decrease energy use and curb global warming. *(Nature's Images)*

OPERATION CHANGE OUT: THE MILITARY CHALLENGE

Operation Change Out: the Military Challenge, a joint project between the DOE and the U.S. Department of Defense (DoD), is the first national military-focused energy-efficiency campaign to encourage every service-man and woman to save energy, money, and protect the environment by replacing their inefficient, incandescent lightbulbs with ENERGY STAR® qualified bulbs. The project was initiated on Earth Day—April 22, 2008.

The goal was to replace at least one incandescent lightbulb with an ENERGY STAR® qualified model in each residential unit at participating military installations.

Even though Operation Change Out began focusing on the U.S. military forces, everyone can participate in the "Change the World, Start with ENERGY STAR" campaign. By visiting www.energystar.gov/changetheworld, it is possible to become a part of a national effort to fight global climate change.

The following tables document the progress so far.

OPERATION CHANGE OUT	
2007–2008	
Number of pledges	42,501
Number of bulbs changed	383,825
Energy savings	108,238,650 kWh
Cost savings	$10,066,194.
Greenhouse gas*	156,984,425 pounds

*pounds of emissions prevented

CUMULATIVE SAVINGS	
Greenhouse gas emissions (lbs)	6,059,599
Dollars	$462,984
kWh	3,656,412
Btus	3,779,901,545

The savings in greenhouse gas emissions and kilowatt-hours makes it worthwhile for everyone to do their part in fighting global warming.

Other energy-efficiency measures can be undertaken as well, such as adequate levels of insulation, tight construction, and high-performance windows and doors. Within office settings, it is important to use energy-saving devices such as sensors that can monitor when a room is empty and can dim the lights. Office equipment designed to be on standby mode when not in use helps cut energy use. Another measure that is convenient and simple to achieve is to purchase recycled office supplies, such as paper. All attempts at conserving energy help the fight against global warming.

GREEN HOMES

According to the EPA, homes can cause twice the greenhouse gas emissions of a car. One way to find out how much energy a home consumes, how to lessen a home's carbon footprint, and how energy and money can be saved is to have a home energy audit completed by a local utility company. An energy audit provides information about how much power a household uses and can supply specific strategies designed to help reduce energy consumption. A home energy audit can help the members of an average household find simple ways to reduce their CO_2 emissions by 1,000 pounds (454 kg) annually and lower utility bills at the same time.

Often, utility companies can perform a thermal scan of a home to detect areas where heat and air may be escaping during the winter or air-conditioned air may be escaping during the summer, causing a home to use much more energy than necessary, which causes electrical generating companies—of which the majority are run by fossil fuels—to send more greenhouse gas emissions into the atmosphere. In a thermal scan, areas that show up bright yellow and orange signify hot areas. During the winter, these are the areas of the home where heated air is escaping. The cooler areas show up in darker tones (blues and purples). These areas signify no energy loss. When areas of energy loss are detected— especially common in older homes—measures can be taken to fix the problems, such as adding better/thicker insulation (for energy loss from the roof) or better windows (another common area of energy loss).

THE ENERGY STAR® PROGRAM

ENERGY STAR® is a joint program of the EPA and the DOE, designed to save money and protect the environment through energy-efficient products and practices.

The detection of wasted heat escaping from a house can be detected through the use of thermal infrared scanners in a branch of science called thermography. The yellows, oranges, and reds (brighter areas) indicate the places where the most energy is escaping. This is usually around the windows and doors of the house. Double-pane insulated windows can reduce energy loss. In homes where heat is escaping through the roof and walls, more insulation is required to prevent energy loss. In homes in London, it is obvious which fireplaces are being used (red)—the blue ones are cold and heat is escaping through most of the windows and doors. *(Wildgoose Education, Ltd.)*

In 1992, the EPA introduced ENERGY STAR® as a voluntary labeling program designed to identify and promote energy-efficient products to reduce greenhouse gas emissions. The first products to receive ENERGY STAR® recognition were computers and monitors. In 1995, the EPA expanded the program to include office equipment.

Today, the ENERGY STAR® label appears on a diverse range of products from major appliances to office equipment to lighting and home electronics.

The program has made significant inroads since its inception through partnerships with more than 12,000 public and private organizations. It educates both organizations and consumers about energy-efficient practices and green living. Recently, the issue has become even more critical as the prices of energy fluctuate wildly in the marketplace. The program provides helpful and informative information on more than 50 product categories involving thousands of different products for homes, schools, hospitals, businesses, and offices. The products listed with an ENERGY STAR® rating are guaranteed to have the same, or better, performance as comparable models while using less energy and saving money.

Geared toward homes and other buildings, ENERGY STAR® saved Americans enough energy in 2007 to avoid greenhouse gas emissions equivalent to those from 27 million cars—a savings equal to $16 billion on utility bills.

Household appliances such as stoves, ovens, refrigerators, freezers, washers, and driers can help save up to one-third off of greenhouse gas emissions and energy bills. New homes are being built across the country certified under the ENERGY STAR® program.

One of the largest benefits of the program is that using energy more efficiently avoids emissions from power plants, avoids the need to construct additional power plants, and reduces energy bills. Because of this, the EPA has determined significant benefits have already been realized. In one year, the ENERGY STAR® program prevented greenhouse gas emissions equivalent to those from 14 million vehicles and avoided using the power that 50 300-megawatt (MW) power plants would have produced, while saving more than $7 billion.

ENERGY STAR® Time Line

The ENERGY STAR® program officially began in 1992 after the inception of the EPA's Green Lights Program. Since then, its influence has grown to touch all aspects of energy issues in the United States, as well as abroad, noted as follows.

1991

- **January:** EPA introduces the Green Lights program, a partnership program designed to promote efficient lighting systems in commercial and industrial buildings (planned to be integrated into ENERGY STAR® by the end of the decade).

1992

- **June:** EPA introduces the first ENERGY STAR®-qualified product line, including personal computers and monitors.

1993

- **January:** ENERGY STAR®-qualified printers are introduced.

1994

- **October:** ENERGY STAR®-qualified fax machines are introduced.

This solar-powered home in Maine generates its own electricity from a 4.25-kW PV system integrated into the rooftop. The south roof incorporates an integrated array of solar thermal collectors and large-area PV modules to form a single, uniform glass pane. Through a net-metering relationship with Central Maine Power, surplus solar electricity is exported to the utility grid, effectively spinning the utility meter backward. Space heating and hot water are provided by the solar thermal system. *(Solar Design Associates, Inc./NREL)*

1995

- **March:** ENERGY STAR® for buildings is launched to help businesses simultaneously improve their energy performance and increase their bottom lines. Green Lights merges with ENERGY STAR® for buildings.
- **April:** EPA introduces ENERGY STAR® specifications for copiers, transformers, and residential heating and cooling products, including air-source heat pumps, central air conditioners, furnaces, gas-fired heat pumps, and programmable thermostats.
- **October:** EPA launches ENERGY STAR®-qualified new homes that are 30 percent more efficient than the model energy code.

1996

- **June:** EPA and DOE announce their ENERGY STAR® partnership. Exit signs, insulation, and boilers are added to the list of qualified product categories.
- **October:** ENERGY STAR® label for appliances, including dishwashers, refrigerators, and room air conditioners is announced.
- **December:** Several national lenders offer ENERGY STAR® mortgages to qualified new home purchases.

1997

- **March:** ENERGY STAR® specifications for residential light fixtures, multifunction devices, and scanners are announced.
- **July:** ENERGY STAR®-qualified clothes washers are announced.
- **December:** ENERGY STAR®-qualified homes expand to include manufactured homes.

1998

- **January:** ENERGY STAR®-qualified TVs and VCRs are announced.
- **March:** ENERGY STAR®-qualified windows are added.

1999

- **January:** ENERGY STAR® program requirements for consumer audio and DVD equipment are announced.

- **February:** ENERGY STAR® label for roof products is introduced.
- **June:** ENERGY STAR® label is extended to office buildings that perform in the top 25 percent of the market.
- **August:** ENERGY STAR®–qualified compact fluorescent lights (CFLs) are added.

2000

- **April:** ENERGY STAR® label is extended to schools that perform in the top 25 percent of the market.
- **June:** U.S. Army and Navy housing procurement specifications comply with ENERGY STAR® qualifications for new homes.
- **October:** ENERGY STAR®–qualified water coolers are introduced.
- **November:** ENERGY STAR®–qualified traffic signals are announced.

2001

- **February:** First home performance with ENERGY STAR® program occurs in New York.
- **July:** ENERGY STAR® label is extended to supermarkets and grocery stores that perform in the top 25 percent of the market.
- **July:** United States and Canada announce an agreement to partner on ENERGY STAR®.
- **September:** ENERGY STAR® specifications for commercial solid door refrigerators and freezers are available.
- **November:** EPA and ENERGY STAR® launch a national public awareness campaign called Change to Encourage Americans to help protect the environment by changing to energy-efficient products and practices.
- **November:** ENERGY STAR® label is extended to acute care hospitals that perform in the top 25 percent of the market.

2002

- **April:** ENERGY STAR® label is extended to hotels that perform in the top 25 percent of the market.
- **May:** More than 100,000 new homes have now earned the ENERGY STAR® label for superior energy performance.

- **June:** First ENERGY STAR® Cool Change campaign is launched.
- **December:** Americans have purchased more than 1 billion ENERGY STAR®–qualified products.
- **December:** Nearly 1,100 buildings have earned the ENERGY STAR® label.

2003

- **January:** ENERGY STAR®, through its work with the U.S. auto manufacturing industry, provides the first plant energy performance indicator for this industry.
- **September:** Cool Change Campaign came to a close. It generated more than $17 million in equivalent and value and more than 70,000 airings in just 18 months.
- **October:** Now 50 percent of the top U.S. homebuilders participate in ENERGY STAR® for new homes.

2004

- **January:** Almost 1,400 buildings have now earned the ENERGY STAR® for superior energy performance.
- **March:** ENERGY STAR® public awareness reaches 56 percent.
- **July:** Updated and expanded agreements with New Zealand on the implementation of ENERGY STAR® are now in place.
- **August:** ENERGY STAR® specification for exit signs is revised.

2005

- **January:** Agreement with the European Union (EU) is updated.
- **January:** Power adapters can now earn the ENERGY STAR® label.
- **January:** Almost 2,000 buildings across the U.S. have now earned the ENERGY STAR® label.
- **February:** ENERGY STAR® public awareness is now over 60 percent nationally.
- **March:** More than 350,000 new homes in the United States have earned the ENERGY STAR®; nearly one in 10 new homes built in 2004 qualified as an ENERGY STAR® home.

- **June:** EPA announces the availability of the ENERGY STAR® performance indicator for auto assembly plants.
- **July:** The Partnership for Home Energy Efficiency (PHEE) program is launched by the EPA, the DOE, and the U.S. Department of Housing and Urban Development (HUD).
- **October:** October 5 is ENERGY STAR® Change a Light Day.
- **December:** Half of the states join the ENERGY STAR® challenge to address energy issues.

2006

- **January:** The EPA adopts guiding principles for designing and operating sustainable federal facilities.
- **January:** More than 2,500 buildings have earned the ENERGY STAR® rating.
- **March:** Americans have purchased more than 2 billion ENERGY STAR®-qualified products.
- **May:** ENERGY STAR® @ Home interactive tool is launched, as part of the annual summer "Cool Your World" campaign.
- **September:** Rebuilt vending machines can now earn the ENERGY STAR® rating.
- **November:** Annual winter Heat Your Home Smartly campaign begins.
- **December:** Nearly 200,000 new homes earned the ENERGY STAR® in 2006 (12 percent of single family home starts), bringing the total number of ENERGY STAR®-qualified homes across the nation to almost 750,000.

2007

- **January:** ENERGY STAR® specification for digital television adapters (DTAs) is announced.
- **February:** More than 3,200 buildings have now earned the ENERGY STAR® label.
- **April:** Public awareness of ENERGY STAR® label exceeds 65 percent.
- **June:** New ENERGY STAR® public service announcement campaign features real people fighting global warming.
- **August:** Report to Congress on server and data center energy efficiency is presented.

- **September:** EPA's online energy rating system for commercial buildings is updated to include greenhouse gas emission factors.
- **October:** Eighth annual ENERGY STAR® Change a Light campaign launches its first national bus tour.
- **October:** The first retail buildings earn the ENERGY STAR® award.

2008

- **January:** ENERGY STAR®–qualified CFL (compact fluorescent lights) sales for 2007 nearly double, reaching a 20 percent market share.
- **February:** The number of commercial buildings and manufacturing plants to earn the ENERGY STAR® increases by more than 25 percent in 2007 to nearly 4,100.
- **April:** EPA launches its Change the World, Start with ENERGY STAR® campaign to help Americans join in the fight against climate change.
- **April:** EPA launches Low Carbon IT campaign, encouraging organizations to enable the power management features on their computers and monitors.
- **April:** National awareness of ENERGY STAR® increases to 70 percent.
- **June:** More than 1,500 organizations now participate in the ENERGY STAR® Challenge.
- **November:** More than 71,000 commercial buildings participate in the ENERGY STAR® program.

2009

- **March:** The EPA and DOE honor 89 businesses and organizations for their outstanding contributions to reducing greenhouse gas emissions through energy efficiency. Awards include the Award for Sustained Excellence, Partner of the Year, and Award for Excellence. Recipients include 3M, the J.C. Penney Company, the Pella Corporation, PepsiCo., Whirlpool, Kimberly-Clark Corporation, Best Buy, and Nashville Area Habitat for Humanity.

- **May:** ENERGY STAR® Indoor air PLUS program begins, designed to protect homes against moisture and mold, pests, combustion gases, and other airborne pollutants.
- **June:** New ENERGY STAR® guidelines are in the process of being revised and updated.

As public awareness continues to increase, the products that the ENERGY STAR® program affects continue to grow. Each additional accomplishment in the program helps the fight against global warming.

According to the EPA, the ENERGY STAR® program has dramatically increased public use and preference of energy-efficient products and practices and is expected to continue to do so in the future. They estimate there are more than 100 million households, which contribute 17 percent of the nation's greenhouse gas emissions, and the program offers potential energy savings in the range of 25 to 30 percent compared with current consumption.

As part of their program, the EPA has put in place what they call their superior energy management criteria, which to date has proven highly successful. They offer the ENERGY STAR® partnership to organizations of all types and sizes. As part of it, senior executives made a commitment to the superior energy management of their buildings or facilities. This top-level organizational commitment has proven to be a catalyst for energy-efficient investments in many of the most successful partner organizations.

Almost 12,000 organizations have partnered with EPA in the pursuit of superior energy management. Partners include more than 425 public organizations such as state and local governments, schools, and universities, more than 880 businesses across the commercial and industrial sectors, and more than 8,000 small businesses.

The EPA will forge partnerships across the commercial and industrial sectors to create and ensure energy efficiency at the top management levels and to facilitate the development of best practices and information sharing. The EPA has already been able to help commercial real estate, public buildings, schools, higher education, health care, hospitality, automobile manufacturing, cement manufacturing, wet corn milling, and others.

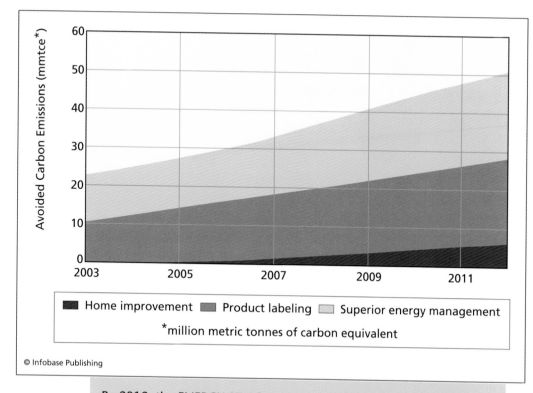

By 2012, the ENERGY STAR® program expects to have potentially eliminated 50 MMTCE of greenhouse gas emissions in an effort to curb global warming.

The future also looks strong for ENERGY STAR®. The EPA expects the program to grow in the future. The following figure illustrates that by 2012 the ENERGY STAR® program is expected to avoid about 50 million metric tons of carbon equivalent (MMTCE) of greenhouse gas emissions each year, equivalent to the emissions from more than 30 million vehicles and reduce energy bills by about $15 billion annually.

Programs like this help educate the public on the realities of environmental health as it relates to energy use and its negative effects on the environment through global warming.

Global Warming and Pollution: Green Transportation

Of all the inventions since the beginning of the industrial revolution, the passenger automobile has had perhaps the most significant impact on society and the environment. After its introduction in 1908 with the Ford Model T, it rapidly gained popularity and became the primary means of transportation in the United States. Today, in the nation, there are about 750 motor vehicles for every 1,000 people. Other countries' growth in number of cars on the road as they become more developed is also growing rapidly. In China, for example, the number of cars has been doubling every five years over the past 30 years.

But for all the freedoms and conveniences cars offer society, there have also been serious consequences—air pollution and energy consumption. Pollution from cars is responsible for respiratory problems, lung cancer, urban smog (ozone), and acid rain. By 1970, the quality of the atmosphere had become so poor that the first federal Clean Air Act was passed in the United States in an effort to control pollution. The act did not prohibit pollution, it just set limits for tolerable amounts. Pollution today, however, is still a huge problem. Other major cities of the world, such as Mexico City, New Delhi, Bangkok, and São Paulo, also fight pollution problems—much of it attributed to automobiles.

This chapter looks at the transportation sector in depth and how energy efficiency, fuel economy, and pollution have affected global warming. It also examines the direction new technology has taken and which cars have been identified as the most environmentally friendly today.

ENERGY EFFICIENCY

Based on studies conducted by the U.S. Department of Energy (DOE), transportation accounts for more than 67 percent of the oil consumed in the United States. The United States imports more than 54 percent of its oil supply. It is estimated that by 2010, the import rate will have risen to 75 percent. The U.S. Federal Highway Administration has determined that the average vehicle on the road today emits more than 600 pounds (272 kg) of air pollution each year. The pollution—carbon monoxide, sulfur dioxide, nitrogen dioxide, and particulate matter—contributes to smog and health problems for many people.

Fortunately, a major goal of cars in today's market is *fuel efficiency*. By being more efficient, less energy is used, adding less to global warming. (See chapter 3 on the Obama administration's new emissions standards. When looking at the efficiency of a car, it is important to understand where energy is being expended. That way, scientists can work to improve areas of energy loss and efficiency.

When fuel is added to the car, not all fuel is converted to energy. In fact, only about 15 percent of the energy from the fuel is actually used to make the car move and run accessories such as the air-conditioning. The rest of the energy—roughly 85 percent—is lost to inefficiencies in the engine and through idling. Because there is such an enormous waste of energy, researchers are busy trying to improve fuel efficiency with advanced technology.

According to the U.S. Environmental Protection Agency (EPA), city driving is one of the most inefficient forms of driving. When cars are stopped at traffic lights idling, about 17.2 percent of the energy is wasted. Technology has been developed to offset this. Integrated starter/generator (ISG) systems help reduce energy losses by automatically turning the engine off when the car is stopped, then restarting it instantaneously when the accelerator is pressed down.

In cars that are gas powered, more than 62.4 percent of the fuel's energy is lost within the internal combustion engine (ICE). ICEs are extremely inefficient at converting the fuel's chemical energy to mechanical energy. Energy is lost to engine friction, when air is pumped into and out of the engine, and wasted when it is converted to heat within the engine. Advanced engine technologies such as variable valve timing and lift, turbo charging, direct fuel injection, and cylinder deactivation can be used to reduce energy losses. Also, diesels are roughly 30 to 35 percent more efficient than gasoline engines, and new advances in diesel technologies and fuels are making these vehicles attractive to many people.

The various accessories in a car—such as air-conditioning, power steering, and windshield wipers—also use up about 2.2 percent of energy. Better fuel economy can be achieved with more efficient alternator systems and power steering pumps. The driveline accounts for about 5.6 percent of the total energy loss. Much of this loss occurs in the transmission. Technologies such as automated manual transmissions (AMT) are currently being developed to correct these losses.

When a car is in motion, it pushes the air in front of it out of the way. The slower the car goes, the less energy is wasted; the faster it goes, the more energy is used. The drag is directly related to the vehicle's shape. Smoother vehicle shapes have already greatly reduced drag (resistance). Even with the progress made to date, however, scientists at the DOE believe further reductions of 20 to 30 percent are possible. Current energy losses through aerodynamic drag are estimated to be 2.6 percent.

Energy is also expended through the braking process—about 5.8 percent. In the physics of forward motion, the car's drive train has to provide enough energy to overcome the car's inertia, which is directly related to its weight. The more a vehicle weighs, the more energy it takes to move it. The car's weight can be reduced by using lighter-weight materials to construct it. Each time the brakes are used, the energy initially used to overcome inertia is lost.

A property called rolling resistance is another way energy is expended—totaling about 4.2 percent. Rolling resistance is a measure of the force necessary to move the tires forward and is directly proportional to the weight of the load supported by the tire. Several technologies have been developed to reduce rolling resistance, such as improved tire tread. For passenger cars, a 5 to 7 percent reduction in rolling resistance increases fuel efficiency by 1 percent.

FUEL ECONOMY

Increasing fuel economy is the best tool available for cutting the nation's oil dependence. Experience has already proven this to be true—fuel economy of cars was doubled between the 1970s and the late 1980s. According to the Union of Concerned Scientists (UCS), the technology needed to increase the average fuel economy of cars and trucks to 40 miles per gallon (MPG) has already been developed—just not implemented. Based on research by the UCS, if the United States increased fuel economy to over 40 MPG over the next 10 years, in 15 years it would have saved more oil than would ever be obtained from the Arctic National Wildlife Refuge (ANWR). The savings from better fuel economy would keep on increasing forever.

There are several ways that fuel economy can be enhanced, and better fuel economy adds less fossil fuel emissions into the atmosphere. One way is driving more efficiently. Aggressive driving—such as speeding, braking fast, and accelerating fast from a standstill—wastes gas. In fact, it can lower gas mileage by 33 percent at highway speeds and 5 percent during city driving.

Observing the speed limit is also important. A vehicle's gas mileage usually decreases rapidly at speeds greater than 60 MPH (97 km/hr). According to the EPA, each 5 MPH a car is driven over 60 MPH is equivalent to paying an additional $.20 per gallon for gas. The estimated fuel economy for observing the speed limit ranges from 7 to 23 percent, depending on the speed driven.

Removing excess weight that the car has to carry also improves efficiency. It is important to avoid keeping unnecessary heavy items in the vehicle. Each extra 100 pounds (45 kg) can reduce the MPG by up to 2 percent. The reduction is based on the ratio of extra weight relative to the vehicle's weight. Because of this, the smaller the vehicle, the more drastic the effect. Avoiding excessive idling improves a vehicle's efficiency. After all, idling gets zero miles per gallon. The larger the engine in the vehicle, the more gas gets wasted.

Using the vehicle's cruise control is another measure that can be used. The EPA has determined that maintaining a constant, steady speed usually conserves gas. Using a car's overdrive gears also increases efficiency because the vehicle's engine speed goes down. It not only saves gas, it reduces engine wear.

Another major way to maximize fuel economy is to keep the car's engine tuned up. The EPA has determined that fixing a car that is out of tune or has failed an emissions test can improve its gas mileage by an average of 4 percent. Fixing serious maintenance problems—such as a faulty oxygen sensor—can improve gas mileage by up to 40 percent. Changes do not need to be major items either. Simple actions like checking and replacing air filters regularly can make a big difference. Replacing a dirty air filter with a clean one can improve gas mileage up to 10 percent.

Inflating tires properly and checking them often are also recommended in order to increase mileage. This simple action can improve gas mileage by 3.3 percent. Underinflated tires can lower gas mileage by 0.4 percent for every 1 pound per square inch (psi) drop in pressure of all four tires.

It is also important to use the recommended grade of motor oil. Using the manufacturer's recommended grade can improve gas mileage by 1 to 2 percent. As an example, using 10W-30 motor oil in an engine designed to use 5W-30 can lower the car's gas mileage by 1 to 2 percent. It is also important to look for motor oil that is labeled energy conserving on the performance symbol because it contains friction-reducing additives.

Planning and combining trips also saves gas. Several short trips taken from a cold start can use twice as much fuel as a longer multipurpose trip. When the engine has already warmed up, it is more efficient. When commuting to work, it is helpful if work hours can be staggered in order to avoid peak rush hours so that a car spends less time sitting in traffic and consuming fuel. If telecommuting (working from home) is

Mass transportation helps lower greenhouse gas emissions, especially with alternative fuel models, such as this hybrid electric bus. *(Nature's Images)*

an option, that is an even better solution—it takes the car completely off the road. Some businesses allow employees to work one or two days a week at home, eliminating some of the time the car is on the road. Some communities offer ride-share (carpooling) programs. This option not only cuts weekly fuel costs but also saves wear and tear on cars. Many urban areas today have special high occupancy vehicle (HOV) lanes on their freeway systems, which are less congested to use.

One long-standing method of saving fuel and combating global warming is through the use of public transit. Mass transportation provides another convenient way to conserve energy. Many modes of mass transit today use innovative and advanced technology designed to be energy efficient.

According to the American Public Transportation Association (APTA), public transportation in the United States saves approximately 1.4 billion gallons (5.3 billion liters) of gasoline and keeps about 1.5 million tons (1.4 million metric tons) of CO_2 out of the atmosphere each year. According to their statistics, however, only 14 million Americans use public transportation on a daily basis; 88 percent of all trips

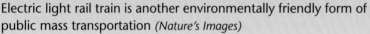

Electric light rail train is another environmentally friendly form of public mass transportation *(Nature's Images)*

in the United States are still made by private automobile—and many of those only carry one person. The APTA has identified the following additional advantages of public transportation.

- Energy independence—if only 10 percent of the American public used mass transportation daily, U.S. dependence on foreign sources of oil would decrease 40 percent.
- Safety—riding a bus is 79 times safer than traveling in a private car; subway and train travel are even safer.
- Cost savings—families that rely on public transportation can reduce their household expenses $6,200 annually—an amount greater than the average family spends on food each year.

Of all the mass transit systems, the train systems are one of the most efficient. They typically emit less carbon and use less fuel per passenger than buses, but are often more expensive to get in place. Electric train systems are even more desirable. Buses that operate on natural gas are also desirable new technology that is spreading rapidly.

Santa Rosa Police Department officer Ken Kimari patrols downtown on his patrol bike. *(Rick Tang, DOE/NREL)*

Bus rapid transit (BRT) is another new alternative that is gaining popularity. This system operates extra-long buses in dedicated lanes. In 2006, a study conducted by the Breakthrough Technologies Institute determined that a BRT system in a medium-sized U.S. city could reduce carbon dioxide (CO_2) emissions by more than 650,000 tons (589,670 metric tons) over a 20-year period. Another option for people inside city limits is walk or ride a bicycle. Some professions such as city law enforcement officers, courier agencies, and delivery services use bicycles as a form of transportation.

CONTRIBUTORS TO GLOBAL WARMING AND POLLUTION

Pollution from cars and trucks is finally receiving attention from the public as global warming is discussed more openly and frequently. The transportation sector is the largest single source of air pollution in the United States today. It causes almost 67 percent of the carbon monoxide (CO), a third of the nitrogen oxides (NO_x), and a fourth of the hydrocarbons in the atmosphere.

Cars and trucks pollute the air during manufacturing, oil refining and distribution, refueling, and, most of all, use. Motor vehicles cause both primary and secondary pollution. Primary pollution is that which is emitted directly into the atmosphere; secondary pollution is from chemical reactions between pollutants in the atmosphere.

The primary ingredient in smog is ozone. Particulate matter consists of soot, metals, and pollen. The finest, smallest textured particulates do the most damage since they travel into the lungs easily. Nitrogen oxides tend to weaken the body's defenses against respiratory infections. Carbon monoxide is formed by the combustion of fossil fuels such as gasoline and is emitted by cars and trucks. When inhaled, it blocks the transport of oxygen to the brain, heart, and other vital organs, making it deadly. Sulfur dioxide is created by the burning of sulfur-containing fuels, especially diesel. It forms fine particles in the atmosphere and is harmful to children and those with asthma. Toxic compounds are chemical compounds emitted by cars, trucks, refineries, and gas pumps and have been related to birth defects, cancer, and other serious illnesses. The EPA estimates that the air toxics emitted from cars and trucks account for half of all cancers caused by air pollution.

According to the UCS, pollution from light trucks is growing quickly. This class of vehicles includes minivans, pickups, and sport utility vehicles (SUVs). Because of the popularity of these vehicles and the extreme number of them on the highways, their emission levels need to be accounted for and controlled. California regulators and the EPA recently created new rules requiring light trucks to become as clean as cars over the next seven to nine years.

Even though for many years there have been air pollution control efforts, 92 million Americans still live in areas with chronic smog problems. The EPA predicts that by 2010, even with current control programs in effect, more than 93 million people will live in areas that violate health standards for ozone (urban smog), and more than 55 million Americans will suffer from unhealthy levels of fine particle pollution.

Trucks and buses are responsible for a large amount of toxic pollution. Although they account for less than 6 percent of the miles driven by highway vehicles in the United States, trucks and buses are responsible for one-fourth of smog-causing pollution from highway vehicles, more than half the soot from highway vehicles, 6 percent of the nation's global warming pollution, and more than one-tenth of America's oil consumption.

Off-highway diesel equipment is another major contributor to pollution. All types of off-highway heavy diesel equipment such as cranes used to build skyscrapers and tractors and combines used in agricultural fields release more fine particulate matter than highway cars and trucks combined. Emissions from this equipment has continued to climb because this equipment has not had to meet the stricter standards that highway vehicles have had to face. Today, a typical tractor emits as much soot as 250 average cars.

NEW TECHNOLOGY

Some cars on the market now offer considerable improvements in fuel economy. Other advanced technologies are under development and will soon be available in new vehicles. Hybrid electric vehicles (HEVs) are also becoming more common on the road. These cars get roughly twice the mileage as conventional vehicles.

The Honda hybrid runs on a combination of gasoline and electric power—hybrids can go more than 50 miles (80.5 km) per gallon of gas and more than 500 miles (804.6 km) on an entire tank. *(Nature's Images)*

The EPA runs a climate change technology program designed to build awareness, expertise, and the capacity to address the risk of climate change at state and local levels. According to the EPA, the transportation sector accounts for 30 percent of U.S. CO_2 emissions from fossil fuel consumption. Roughly 67 percent of these emissions are from gasoline consumption in cars and other vehicles on the roads. The rest comes from other transportation activities, such as diesel-fueled heavy-duty vehicles and jet-fueled aircraft. Cars and trucks alone account for nearly half of all air pollution in the United States and more than 80 percent of urban air pollution.

Concerns about air pollution, energy security, and global warming have encouraged the development of alternative fueled vehicles (AFVs) and policies to encourage their use by the EPA. In fact, the EPA has identified the following benefits associated with AFVs:

- reduced dependence on foreign oil
- job creation
- less air pollution and fewer emissions of greenhouse gases
- potential for reduced fuel and maintenance costs
- positive economic impacts, particularly with *alternative fuels* derived from domestic resources

Alternative fuels not only burn cleaner—producing lower emissions—but some are even renewable (unlike fossil fuels), which means a continuous supply could be developed. AFVs run on other fuels such as compressed natural gas, ethanol, methanol, biodiesel, hydrogen, propane, and electricity. Specific characteristics of these fuels will be discussed further in chapter 8.

In addition to the fact that AFVs improve urban air quality, they also help cut back on greenhouse gas emissions. For example, compressed natural gas (CNG), liquid petroleum gas (LPG), and corn-based ethanol emit less CO_2 than gasoline does if the full fuel cycle is considered. Fuel cells and electric vehicles also have the potential to reduce greenhouse gas emissions significantly. Even though the initial purchase prices of these vehicles may be higher right now than traditional fossil fuel–burning vehicles, some AFVs, such as electric and natural gas vehicles, have lower fuel and maintenance costs than gasoline vehicles do. In addition, the federal government and some state governments offer tax incentives and grant programs to improve the affordability of AFVs. Some automakers are even offering rebates and other incentives to make these cars affordable and accessible. Many automakers are currently developing fuel cell vehicles, which will use methanol or hydrogen as fuel.

There is also federal government activity and backing behind AFVs. The Energy Policy Act of 1992 (EPACT) requires federal, state, and fuel provider fleets to acquire alternative fueled vehicles. It will require private and local government fleets to acquire AFVs. Under the Clear Air Act amendments of 1990, fleet vehicles in places with high levels of air pollution must use alternative fuels, reformulated gasoline, or clean diesel fuel.

The Clean Cities program, coordinated by the DOE, is a locally based partnership of government and industry to expand the use of alternative fuels by accelerating the deployment of AFVs and building a local refueling infrastructure. Currently, there are 75 Clean Cities around the country, and the 3,500 stakeholders own and operate 150,000 AFVs.

According to an article in the *New York Times* about a new AFV being developed, "It looks like an ordinary family sedan, costs more to build than a Ferrari, and may have just moved the world one step closer to a future free of petroleum."

In "Latest Honda Runs on Hydrogen, Not Petroleum," the piece which appeared on June 17, 2008, introduced the new invention as Honda Motor's FCX Clarity—the world's first hydrogen-powered fuel cell vehicle intended for mass production.

Over the next three years, Honda will produce only 200 of the futuristic-looking FCX Claritys. It then will increase production as cities gear up to provide fueling stations accessible across the country.

According to Kazuaki Umezu, head of Honda's Automobile New Model Center, "Basically, we can mass produce these now. We are waiting for the infrastructure to catch up."

The attractiveness of fuel cell vehicles is that their only emissions are water and heat—there is no air-polluting exhaust. Fuel cells work by com-

The Sacramento Municipal Utility District's solar-powered hydrogen vehicle fueling station. As the solar panels make electricity, an electrolyzer at the station uses that energy to separate water into hydrogen to make fuel for hydrogen-powered vehicles. When no energy is being used to produce hydrogen for vehicles, the power produced by the panels goes into the Sacramento Municipal Utility's grid. *(Keith Wipke, DOE/NREL)*

bining hydrogen and oxygen from ordinary air to make electricity. Takeo Fukui, Honda's president, said, "This is a must-have technology for the future of the Earth. Honda will work hard to mainstream fuel cell cars."

Fuel cells are seen as an advantage over electric cars, whose batteries take hours to recharge and use electricity, which (especially in the United States and China) is often produced by coal-burning power plants.

Honda says the FCX Clarity can be filled easily at a pump and can drive 280 miles (451 km) on a tank. It also gets higher fuel efficiency than a gasoline car or hybrid—the equivalent of 74 MPG. To date, the technology has faced hurdles, such as the high price to acquire fuel cells, but Honda claims it has found ways to produce them, which they expect to significantly lower the costs. Fukui says the cars cost several hundred thousand dollars each to produce, yet that cost should drop below $100,000 in less than a decade as production volumes increase. According to the article, fuel cell vehicles have been a big gamble for Honda—they have spent the past 16 years and millions of dollars in research and have been criticized for not branching out into the SUV market like everyone else.

Fuel cells have come down in size and in the FCX Clarity they fit in a box the size of a desktop computer that weighs about 150 pounds (68 kg), which is less than half of their size just 10 years ago. The FCX Clarity's fuel cell unit can generate up to 100 kilowatts of electricity—enough to accelerate the car from 0 to 60 MPH (0–97 km/hr) in less than nine seconds and give it top speeds of 100 miles (161 km) an hour. Honda claims, "The FCX Clarity looks like a four-door sedan; looks like a sleeker version of the Accord, and drives with the hushed whine of a golf cart."

Honda says a big remaining hurdle to mass production is the lack of filling stations that sell hydrogen. Even in California, where the state government has promoted and encouraged the construction of hydrogen stations, there still are not enough.

IN THE NEWS—THE TOP ENVIRONMENTALLY FRIENDLY CARS

In light of global warming and other environmental issues, automobile manufacturers worldwide have been working to make vehicles more fuel efficient and environmentally friendly. Research conducted by J. D. Power and Associates in California and compiled in its automotive environmental index (AEI) listed the top environmentally friendly vehicles chosen as a result of their research. Their study took into account information supplied by the EPA and consumers related to fuel economy, air pollution, and greenhouse gases.

Mike Marshall, director of automotive emerging technologies, said, "High gas prices, coupled with consumers becoming more familiar with alternative powertrain technology, are definitely increasing consumer interest in hybrids and flexible fuels. However, the additional price premiums associated with *hybrid vehicles,* which can run from $3,000 to $10,000 more than a comparable non-hybrid vehicle, remain the biggest concern among consumers considering a hybrid. The AEI highlights several non-hybrid models available that help consumers reduce fuel use and emissions."

Based on the study, there is high consumer interest in hybrids and vehicles that run on alternative fuels such as diesel or E85 (ethanol). According to the study, less than one-fourth of consumers questioned will only consider a gasoline-powered car for their next purchase—75 percent were interested in an alternative fuel vehicle. According to Marshall: "There is a real need to educate consumers about the technology and its benefits."

At the 2009 Detroit Auto Show, the focus was on "reducing harmful greenhouse gases and U.S. dependence on foreign oil . . . two of the biggest hurdles facing the nation today."

30 Most Environmentally Friendly Cars		
Acura RSX	Chevrolet Aveo	Chevrolet Cobalt
Ford Escape Hybrid	Ford Focus	Ford Focus Wagon
Honda Accord	Honda Accord Hybrid	Honda Civic
Honda Civic Hybrid	Honda Insight	Hyundai Accent
Hyundai Elantra	Kia Rio	Kia Spectra
Lexus RX400h	Mazda Mazda3	Mazda MX-5 Miata
Mercury Mariner Hybrid	Nissan Sentra	Saturn Ion
Scion xA	Suzuki Reno	Toyota Camry
Toyota Corolla	Toyota Highlander Hybrid	Toyota Prius
Volkswagen Golf	Volkswagen Jetta	Volkswagen New Beetle

Accomplishing both tasks will require the same course of action—reducing the amount of fossil fuel that is burned every day. According to the National Resources Defense Council (NRDC), America spends more than $200,000 per minute on foreign oil—$13 million per hour. Two-thirds of that is used for transportation. By increasing the efficiency of cars, trucks, and SUVs, many of the environmental problems can be solved. Driving more fuel-efficient cars—fuel sippers, not guzzlers—is one way to achieve this. The 2009 Detroit top fuel-sipper picks are as follows:

- 2010 Audi A3 2.0 TDI; 40+ MPG
- Chrysler 200C EV; no gas—battery only
- Dodge Circuit; no gas—battery only
- 2010 Ford Fusion Hybrid; 40 MPG (700 miles/tank)
- 2010 Honda Insight; 43 MPG
- 2010 Lexus HS250h; 35+ MPG
- Mercedes-Benz Blue ZERO; Electric
- 2010 Toyota Prius; 50 MPG
- Volkswagen Blue Sport; 55 MPG
- 2009 Volkswagen Jetta TDI SportWagon; 30+ MPG

To date, Honda is known as the greenest automaker. They have the best overall smog performance in four out of five classes of vehicles and better-than-average global warming scores in every class. As more consumers become aware and educated, automobile manufacturers will be pressured into meeting the demands for greener technology.

Global Warming and Pollution: Cities and Industry

Cheap and abundant fossil fuels have created bad energy habits, especially in the most wealthy, developed countries. Two prominent sectors that contribute to global warming are cities and industry, which have a responsibility to the environment to do their part in cutting back on emissions—either through the employment of different practices or the implementation of new technology—in order to help solve the global warming problem.

This chapter looks first at the major issues cities face with solid waste and how that affects greenhouse gas emissions, landfills, and opportunities for methane recovery, and then at ways cities can empower themselves to help solve problems by taking positive action. Next, it explores the industrial sector, power generation, pollution, and clean coal technology.

GLOBAL WARMING AND CITIES

One of the major issues in cities is the maintenance and disposal of solid waste. The disposal of solid waste contributes to greenhouse gas

(GHG) emissions in many ways. Because of this, it is important that cities address not only how and which GHG are produced, but ways the emissions can be reduced and controlled as well.

As shown in the following illustration, the sources of GHG emissions come from several phases that a product goes through during its life cycle, from raw material acquisition, to the manufacturing process, to its use, to its final disposal (whether it is recycled and used again, composted, burned, or stored in a landfill). Each phase has specific processes associated with it, and each process can have specific GHG emissions associated with it. For example, the manufacturing process often releases carbon dioxide (CO_2) into the atmosphere. In the case of landfills, anaerobic decomposition of waste produces methane—a GHG 21 times more potent than CO_2. When waste is incinerated as part of a waste management system, CO_2 is produced as a by-product. When waste is transported from a collection site to a disposal site, the transportation vehicle produces CO_2 emissions from its use of fossil fuels. For waste that is deposited in landfills, completely new products must then be manufactured to replace the item that has just been thrown away. This takes energy to create the product and adds any GHG emissions to the atmosphere that are generated through the acquisition of the raw materials and manufacture of the product. Had the product been recycled, the acquisition of raw materials could have been avoided, eliminating those GHG emissions, and possibly some of the manufacturing process could have been avoided as well.

According to the U.S. Environmental Protection Agency (EPA), waste prevention and recycling—referred to collectively as waste reduction—help better manage solid waste. Even better strategies include preventing waste altogether or recycling. If waste is prevented where possible and recycled where not, then methane emissions can be reduced at landfills. As shown in the illustration, the four main stages of product life cycles, which all provide opportunities for GHG emissions and/or offsets, are raw material acquisition, manufacturing, recycling, and waste management.

Composting releases some CO_2 into the atmosphere. Combustion releases both CO_2 and nitrous oxide. Nitrous oxide is 310 times more

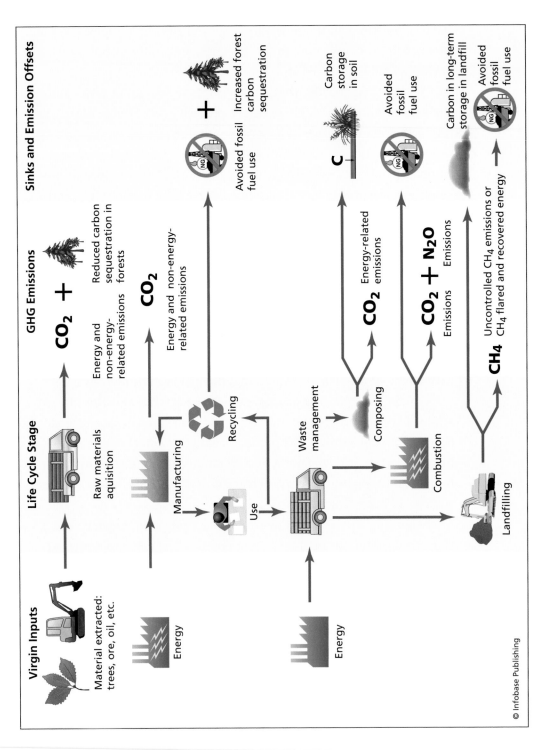

potent than CO_2. Landfilling is the most common whole management practice. One of the biggest problems with landfills, other than the space they take up, is the fact that as the organic material decomposes under anaerobic conditions (without the presence of oxygen) it produces methane.

Carbon-containing materials contained in landfills that do not decompose fully sequester that carbon during the lifetime the material remains buried in the landfill, keeping it from being released into the atmosphere. One promising development concerning landfills is that modern technology has developed to the point that landfill methane can now be captured and converted into a source of energy. According to the EPA, landfills are the single largest human source of methane emissions in the United States. It is produced by the bacterial decomposition of organic materials such as yard waste (grass clippings, weeds, etc.), household waste, food waste, and paper. Methane actually creates an explosion hazard in landfills. Landfill gas also contains volatile organic compounds that contribute to ground-level ozone (smog).

Today, the Clean Air Act requires many landfills to collect and burn their landfill gas emissions. Energy recovery can use the energy value of landfill gas and displace the use of fossil fuels. With this technology, offsetting the use of coal and oil to generate electricity or heat reduces emissions of GHGs and other pollutants (such as sulfur dioxide—a major contributor to acid rain).

Another positive feature of landfill gas is that it is constantly generated, enabling it to be a reliable fuel for several energy applications, such as power generation or direct use. Electric utilities can use landfill gas-to-energy renewable energy projects—a viable source of green power. According to the EPA, it is even feasible for industrial facilities, uni-

(opposite page) The four main stages of product life cycles, all of which provide opportunities for GHG emissions and/or offsets, are raw material acquisition, manufacturing, recycling, and waste management.

versities, hospitals, and other large energy users to benefit by connecting directly into landfill gas from local landfills once facilities are set in place. Once connected, these large operations can burn the landfill gas to provide their own heat, hot water, or electricity.

Concerning the waste that is already in landfills, the EPA has a landfill methane outreach program (LMOP) that puts the waste to a good use. As the organic wastes within a landfill decompose, they produce methane gas—a GHG that contributes to global warming. The LMOP shows communities and companies how to capture landfill gas and convert it to energy.

Currently, LMOP is a voluntary assistance and partnership program that promotes the use of landfill gas as a renewable, green energy source. The gas is composed primarily of CO_2 and methane. The LMOP forms partnerships with communities, landfill owners, utility companies, power marketers, states, tribes, project developers, and nonprofit organizations. The EPA set this program up as part of the U.S. commitment to reduce greenhouse gas emissions under the United Nations Framework Convention on Climate Change (UNFCCC).

Landfill gas (LFG) is extracted from landfills using a series of wells and a blower/flare (or vacuum) system. The system carries the collected gas to a central point where it is processed. From there, it can either be flared or used to generate electricity, replace fossil fuels in industrial and manufacturing operations, fuel greenhouse operations, or be upgraded to pipeline quality gas. As of December 2007, there were approximately 445 operational LFG energy projects in the United States. There are several methods for converting LFG to energy—electricity generation, direct use, cogeneration, and alternate fuels.

Electricity Generation

Electricity can be generated for on-site use or for sale to a power grid. Different technologies can be used to generate the electricity, including internal combustion engines, turbines, microturbines, stirling engines (external combustion engines), *organic Rankine cycle* engines, and fuel cells.

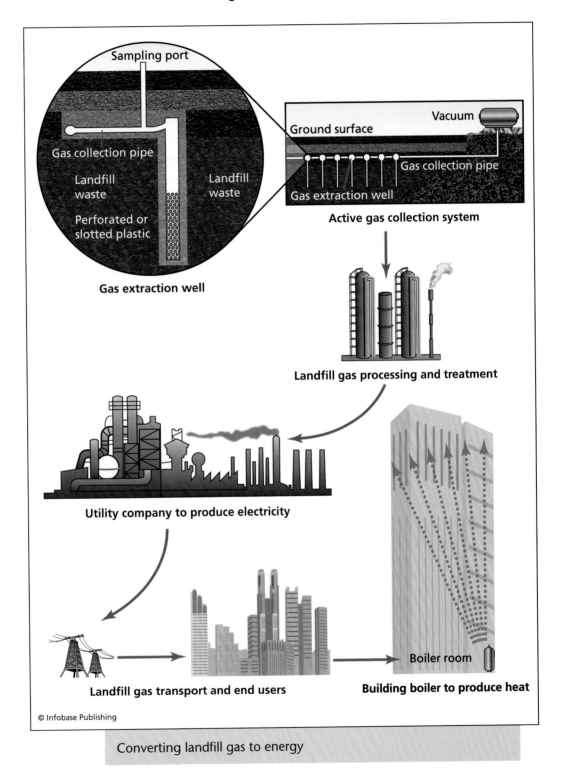

Sampling port

Gas collection pipe

Landfill waste

Landfill waste

Perforated or slotted plastic

Gas extraction well

Vacuum

Ground surface

Gas collection pipe

Gas extraction well

Active gas collection system

Landfill gas processing and treatment

Utility company to produce electricity

Landfill gas transport and end users

Boiler room

Building boiler to produce heat

© Infobase Publishing

Converting landfill gas to energy

Direct Use

Almost one-third of the projects utilize the direct-use method. This option directly uses LFG to offset the use of another fuel, such as natural gas, coal, or fuel oil. The fuel can be used in a boiler, kiln, greenhouse, dryer, or other thermal applications. Currently, it is being used for applications such as firing pottery and glass-blowing kilns, powering and heating ice rinks and greenhouses, and heating water for aquaculture (fish farming) operations. Current industries using LFG include chemical production, auto manufacturing, food processing, cement and brick manufacturing, the pharmaceutical industry, and prisons and hospitals.

Cogeneration and Alternate Fuels

Cogeneration uses LFG to generate both electricity and thermal energy, usually in the form of steam or hot water. The production of alternative fuels is still in the early stages of development. According to the EPA, landfill gas has been successfully delivered to the natural gas pipeline system as both a high-Btu and medium-Btu fuel. It has also been converted to vehicle fuel as compressed natural gas (CNG) and liquefied natural gas (LNG). Projects to convert LFG to methanol are in the planning stages.

There are several items of practical action that individual cities can take in order to fight climate change and global warming. The EPA has conducted studies across the nation and developed a list of 20 municipal-level actions that the public and businesses can take that can save money, save energy, keep the air clean, reduce congestion, curb and minimize urban sprawl, and reduce greenhouse gas emissions. They have recommended that the public take it upon themselves to do the following:

1. Make building energy improvements

   Municipal buildings can make a significant difference when they remodel to install energy-efficient alternative items.

   Example: The EPA reports that municipal building retrofits in Chicago are reducing CO_2 emissions by 7,602 tons

(6,896 metric tons) a year and saving the city budget almost $1 million annually.

2. Replace motors in city operations with more efficient models

Energy-efficient motors can cut energy consumption, reduce GHG emissions and other air pollution, and save money.

Example: Long Beach, California, improved the efficiency of its recycling and solid waste-to-energy plant by installing variable frequency drives, saving $329,508 per year in energy costs and reducing annual CO_2 emissions by more than 3 million pounds (1.4 million kg).

3. Purchase ENERGY STAR® equipment for municipal offices

Copiers, computers, scanners, fax machines, heating and cooling products, windows, and other equipment with the ENERGY STAR® label save money and reduce GHG emissions.

Example: Portland, Oregon, is using ENERGY STAR® office equipment in all its city purchases now. The city operates more than 2,000 computers. If 30 to 40 percent of the users leave their equipment running 24 hours a day and on weekends, Portland's savings from using ENERGY STAR® equipment could still approach $35,000 each year.

4. Change traffic lights to light-emitting diode (LED) fixtures

LEDs are 80 to 90 percent more efficient and last 10 times longer than ordinary lights, which reduces energy and maintenance costs.

Example: Saint Paul, Minnesota, installed red LEDs and red arrows at more than 200 intersections citywide and have projected an annual savings of more than $135,000 and 1,250 tons (1,134 metric tons) of CO_2.

5. Use renewable energy systems to improve air quality

Switching from fossil fuel–generated electricity to renewable-based power is a smart way to reduce GHG emissions.

Example: In Austin, Texas, they plan to meet 50 percent of all new electricity demand with renewable energy by

2010. Achieving this goal would reduce CO_2 emissions by 1.9 million tons (1.7 million metric tons) per year.

6. Purchase green power to improve air quality

In states that have competitive electricity markets, there are often options to purchase shares of green energy.

Example: Santa Monica, California, purchased five megawatts of green electricity to power all of its municipal facilities. The $2.3 million that the city spends annually for electricity will go to companies that contract directly with renewable generators.

7. Redesign communities to encourage walking, biking, and mass transit

Every gallon of gas burned by a vehicle releases 20 pounds (9 kg) of CO_2 into the atmosphere, and vehicles are major contributors to urban air pollution.

Example: Xenia and Green County, Ohio, took 60 miles (97 km) of former railway corridors and an old railway depot and converted them into an attractive transportation center and route. It is now used for pedestrians and bicyclists. Thousands of people use these routes now to go to work, school, and other destinations.

8. Provide incentives for mass transit and carpooling

This option leaves the creativity up to individual businesses to make it worth their employees' while.

Example: West Hollywood, California, has a parking cash-out program in which city hall employees receive cash incentives of up to $65 each month to leave their cars at home and use alternative methods to commute to work.

9. Foster telecommuting and similar trip reduction programs

Working at home or at telecommuting centers reduces vehicle miles traveled.

Example: San Francisco's public defenders office conducts interviews with inmates at two county jail facilities in San Bruno, California. This program has eliminated the need for a 40-mile (64-km) round-trip between facilities and reduced annual vehicle miles traveled by 600,000 (966,000 km) and annual CO_2 emissions by 351 tons (318 metric tons).

10. Convert fleets to run on alternative fuels

    Using fleets that run on compressed gas, ethanol, methanol, biodiesel, hydrogen, and electricity can improve urban air quality and reduce greenhouse gases.

    Example: Chattanooga, Tennessee, provides free electric bus service in their downtown area. The shuttle system eliminates 3.5 million pounds (1.6 million kg) per year of CO_2 emissions.

11. Put police on bicycles

    Running bicycle programs saves on fuel, vehicle and maintenance costs, and can improve the police departments' ability to serve and protect citizens.

    Example: Dayton, Ohio, saves $27,000 each year in reduced fuel and maintenance costs and reduces CO_2 emissions from police transportation by 7.5 tons (6.8 metric tons) per year.

12. Initiate pay as you throw waste disposal programs

    Charging residents for the collection of household trash based on the amount they throw away creates a direct economic incentive to recycle more and generate less waste.

    Example: From 1990 to 1995, Mount Vernon, Iowa's pay-as-you-throw program cut the amount of trash sent to landfills by 40 percent, almost doubled the recycling rate, and virtually eliminated disposal of yard waste.

13. Implement curbside recycling

    Recycling saves energy by reducing the fossil fuels needed to extract and manufacture new products and, in the case of paper products, increases carbon sequestered in forests.

    Example: In Hillsborough County, Florida, nearly 800,000 tons (725,748 metric tons) of CO_2 equivalent are avoided each year through waste reduction, recycling, and composting programs.

14. Recycle office paper and reduce landfill costs

    Recycling reduces the energy and materials needed to produce new paper.

    Example: In 1998, Littleton, New Hampshire, recycled 68 tons (62 metric tons) of mixed office paper. If the town had thrown that paper away, disposal and trucking fees would have cost $3,876. The cost for recycling was $1,020. The town ended up selling the paper for $5,950 to a company that produces toilet tissue and paper towels.

15. Buy products made from recycled materials

Recycled products generally require less energy to produce than new products; many recycled products cost less than new ones.

Example: In 1998, Metropolitan King County, Washington, saved about $600,000 by purchasing recycled materials such as toner cartridges (saved $300,000), retreaded tires (saved $77,000), and shredded wood-waste for road surfaces, landscaping, and erosion control (saved $65,000).

16. Establish composting programs

Composting organic wastes reduces methane emissions and keeps waste out of landfills.

Example: In Albuquerque, New Mexico, a green waste composting program composts yard and stable waste. The end product is marketed through a local garden center. The program diverts 9,570 tons (8,682 metric tons) of waste from the landfill and reduces GHG emissions by 4,626 tons (4,197 metric tons) of CO_2 equivalent each year.

17. Capture methane from landfills

Decomposing trash in landfills produces landfill gas—about 50 percent methane.

Example: Prince George's County, Maryland, installed a methane recovery system at a landfill and uses the methane to provide heat, hot water, and electricity to a nearby correctional facility. The county then sells the leftover methane to a utility company. Annual energy revenues are about $1.3 million, and methane emissions have been reduced by 45,000 tons (40,823 metric tons)—a GHG reduction equivalent to that achieved by planting almost 83,000 acres (33,589 hectares) of trees.

18. Integrate smart growth in planning

This is metropolitan development that pays for itself while protecting air and water quality, encouraging redevelopment of former industrial sites, and promoting community economic ventures.

Example: Portland, Oregon, practices smart growth by increasing the use of land within its urban growth boundary and redeveloping. One project is currently creating 5,700 new jobs.

19. Plant trees to keep buildings and streets cooler and improve air quality, lower air-conditioning loads, and save money

Trees provide shade, reducing the amount of energy needed to cool buildings.

Example: In Miami, Florida, the city planted 88 trees distributed among 14 homes. Each homeowner has saved $20 per year in energy costs, and the project reduced the neighborhood's annual energy-related CO_2 emissions by 42.13 tons (38.22 metric tons) while storing 0.33 tons (0.30 metric tons) per year in the growing trees.

20. Use highly reflective surfacing and roofing materials

Highly reflective roofs and pavements help make cities cooler, reduce the formation of smog, reduce air-conditioning loads, and save money. They can reduce air-conditioning expenses by 10 to 50 percent.

Example: Frederick, Maryland, saves about $1 million each year in cooling costs from its existing highly reflective roofs and tree plantings.

By taking positive action, it is possible to lower greenhouse gas emissions and save energy, while lowering expenses and dependence on foreign oil.

GLOBAL WARMING AND INDUSTRY

Industry is a major contributor to global warming. Through the burning of fossil fuels, large amounts of CO_2 and particulates are being added to the atmosphere on a daily basis. Power generation is one of the biggest offenders among industries. This section looks at power generation, soot and particle pollution, and clean coal technology.

Power Generation

Power plants generally only turn about 30 percent of the energy input into usable electricity. Up to 75 percent of the energy is lost in the fuel at the start of the process. Wasting energy is a common occurrence. If energy companies combined power and heat production (a concept called cogeneration), they could increase efficiency to over 70 percent. One of the biggest problems is that many power-producing plants are operating with older equipment.

According to the American Association for the Advancement of Science, human industrial activity is largely responsible for the current global warming situation. Records show that over the past 50 years the warming trend has sped up largely because of the greenhouse gases being added to the atmosphere by a wide range of industrial activities—everything from power plants to the production of gas-guzzling cars.

Large, central-station power plants are often powered by coal, causing greenhouse gas emissions to enter the atmosphere from the burning of fossil fuels. *(Warren Gretz, DOE/NREL)*

Tim Barnett of the Scripps Institution of Oceanography, commenting on the steadily rising temperature, said, "We were stunned by the similarities between the observations that have been recorded at sea worldwide and the models that climatologists made. The debate is over, at least for rational people. And for those who insist that the uncertainties remain too great, their argument is no longer tenable. We've nailed it.

"For the past 40 years, observations by seaborne instruments have shown that the increased warming has penetrated the oceans of the world—observations that have proved identical to computer predictions whose accuracy has been challenged by global warming skeptics. The most recent temperature observations fit those models with extraordinary accuracy."

One of the sectors most responsible for global warming is the coal-fired power plant. Many refer to this common source of power as dirty coal power. More than half of the electricity generated in the United States is generated from coal-fired plants—unfortunately, one of the dirtiest sources of electricity.

When the Clean Air Act was passed in the 1970s, Congress included a grandfathering loophole that allowed older power plants to be exempt from meeting the new stricter regulations that newer plants would be required to meet. Congress allowed the loophole because they figured all of the older plants would eventually be replaced by newer, conformable plants. But that is not what happened. In fact, most existing power plants today are between 30 and 50 years old and up to 10 times dirtier than new power plants.

Today, power plants are a major source of pollution, with coal-fired power plants spewing 59 percent of total U.S. sulfur dioxide pollution and 18 percent of total nitrogen oxide pollution into the atmosphere every year. These power plants can also release over 40 percent of total U.S. CO_2 emissions.

In a study conducted by the Massachusetts Institute of Technology, researchers claimed that under the most optimistic scenario GHG emissions that contribute to global warming can be stable by 2050 at 2000 levels, with nuclear power and renewable sources replacing or helping some of the coal-fired generating capacity.

Soot and Particle Pollution

According to *National Geographic,* soot (a prime component of city smog) is the second-largest contributor to greenhouse gases in the atmosphere. Mark Jacobson, an assistant professor of civil and environmental engineering at Stanford University, says, "Soot—or black carbon—may be responsible for 15–30 percent of global warming. Soot is comprised of fine black carbon particles formed by incomplete combustion. It generally originates in the atmosphere as a by-product of coal-burning power plants; diesel-burning cars, trucks, buses, and tractors; jet fuel; forest fires; wood burning stoves and fireplaces; dung-fueled fires for heat and cooking; kerosene; and candles." According to Jacobson, "The lifetime of soot is anywhere from a week to a few weeks. And the thinking was that they didn't have time to combine with other particle types in that relatively short lifetime."

Scientists believed that the particles, once emitted into the atmosphere, floated suspended in the atmosphere until they fell back to Earth under their own weight or were washed out of the atmosphere by rain. But Jacobson, who specializes in computer modeling of atmospheric pollution, developed a program to determine if, and how much, soot combines with other particles in the atmosphere. The model showed that within five days after entering the atmosphere, particles of pure soot could be found in mixtures containing dust, sea spray, sulfate, and other chemicals.

According to Jacobson, "Identifying the types of pollution that contribute to climate change is the essential first step in adopting rational conservation initiatives."

What was important about Jacobson's model was that prior to his discovery, the assumption was that soot was not combining with other particles and chemicals once it reached the atmosphere, so it had not been dealt with in any international discussions about controlling climate change. The Intergovernmental Panel on Climate Change (IPCC), for example, focused primarily on CO_2 and methane—not soot. Kyoto also focused on greenhouse gases, not soot. Jacobson stresses that soot needs to be looked at too. The control and elimination of soot is just as critical. He argues that policy makers need to consider soot just as important as GHG.

Pollution from a power plant using coal to generate electricity
(DOE/NREL)

According to Jacobson: "How do we reduce soot emissions? That's an easy question. Get rid of a lot of old cars . . . in countries other than the United States and some in Europe the technology to control emissions is not being used. Tighten standards and the technology will follow. After all, the catalytic converter [that controls car emissions] wasn't developed until after the Clean Air Act was passed in the 1970s.

"The technology exists to retrofit older coal-fired power plants. Most important is phasing out dirty fuels like coal and diesel, and replacing them with alternative energies. Wind energy is selling for 3.5 cents per kilowatt-hour, which is certainly affordable. But right now Germany has more wind power than the United States. The U.S. can do a lot more."

According to a report on the Web site LiveScience, using computer models and information from NASA satellites, scientists have located significant accumulations of black carbon soot in the Arctic, which may be contributing to the warming of that region that has already experienced a significant rise in temperature and melting of ice.

Dorothy Koch, of Columbia University and NASA Goddard Institute for Space Studies (GISS), said, "This research offers additional evidence that black carbon, generated through the process of incomplete combustion, may have a significant warming impact on the Arctic."

This is happening because dark, dirty ice absorbs sunlight and traps heat. Airborne soot also traps heat, causing warming air and altering weather patterns and cloud cover. Of the atmospheric soot found above the Arctic, roughly one-third originates from south Asia, which has the largest industrial soot emissions in the world. Russia, North America, and Europe are also significant industrial producers of soot. One-third of all soot comes from the worldwide burning of trees and biomass.

According to an article in the journal *Nature*, pollution-filled brown clouds also enhance solar heating of the lower atmosphere by about 50 percent. Research done at the Scripps Institution of Oceanography has determined that the combined heating effect of greenhouse gases and brown clouds, which contain soot, trace metals, and other particles from urban, industrial, and agricultural sources, is enough to account for the retreat of the Himalayan glaciers during the past 50 years.

The atmospheric physicist Veerabhadran Ramanathan of Scripps said, "If it becomes widespread and continues for several more decades, the rapid melting of these glaciers, the third-largest ice mass on the planet, will have unprecedented effects on southern and eastern Asia."

Another interesting development is that some coal-fired plant construction proposals around the country are beginning to be met with disapproval because of their contribution to global warming. The Kansas Department of Health and Environment said that CO_2 emissions from two proposed 700-megawatt coal-fueled units would threaten public health and the environment. The agency rejected an air permit for the permit. Roderick Bremby, the department's secretary, said "It would be irresponsible to ignore emerging information about the contribution of CO_2 and other greenhouse gases to climate change and the potential harm to our environment and health if we do nothing."

Air permits have been denied before, but this is the first incident where CO_2 has been the basis for rejection. Nationwide, at least 16 coal-fired power plant proposals have been recently dropped and more than three dozen delayed. According to the U.S. Department of Energy

INDUSTRY ARGUMENTS AGAINST GLOBAL WARMING ACTION AND WHY THEY ARE WRONG

The following arguments are what policy makers, businesses, scientists, and others often hear from industries when they are urged to take positive action to clean up industrial practices and curb global warming emissions. Their arguments are followed by the response of scientific specialists involved in the issue.

1. Argument: Binding emissions reductions before 2020 are too swift and should not be imposed until the technology to remove CO_2 from coal-fired power plants is commercially available.

   Response: Global warming is under way. Severe swings in regional climates are already causing natural disasters to sweep the globe. Global warming poses a major threat to international stability. The IPCC says we must begin reductions by 2015 to have any realistic chance to prevent the worst effects of global warming. The technology exists to reduce these effects through comprehensive application of energy efficiency, wind power, and solar power. New coal-fired power plants should also prepare to make CO_2 reductions right away. Carbon capture and sequestration (CCS) technology will be commercially available between 2015 and 2020. If the first wave of near-zero emissions is expected to start operation around 2012 to 2013, then around 2015 commercial availability of CCS technologies should be available for new plants and retrofit of some existing plants.

2. Argument: Global warming reductions will drive oil and gasoline prices even higher.

   Response: High-energy prices are the result of status quo energy policies that rely heavily on a single fuel—oil—for a single use—transportation. Oil companies have seen record profits under the current approach—more than $650 billion from 2001 to the first quarter of 2008. In the first half of 2008, oil prices increased by 44 percent.

3. Argument: Global warming reductions will decimate families' budgets.

(continues)

(continued)

Response: Americans' budgets are already under siege from higher food and fuel costs and stagnant wages. The decrease in median family income puts working families into even more of a bind. Meanwhile, their hard-earned dollars help oil companies make record-setting profits.

4. Argument: Global warming reductions will send American jobs overseas to countries that do not reduce their emissions.

    Response: Investments in clean energy, boosted by binding reductions in global warming pollution, would generate millions of new jobs. Today, annual investments in energy-efficiency technologies currently support 1.6 million U.S. jobs. Renewable energy such as wind and solar power created 450,000 new jobs in 2006 alone. The Climate Security Act would create tens of thousands of new clean-energy jobs.

5. Argument: The Climate Security Act will wreck the economy.

    Response: The Climate Security Act would have almost no negative effect on long-term economic growth, according to a number of economic models.

6. Argument: Global warming solutions will hurt the poor.

    Response: Low-income households are already harmed by high energy prices. More of the same energy policies will only compound these problems. The Climate Security Act would provide more than $1 trillion to assist such families via a tax cut, clean-energy job training, and rate relief to protect consumers from higher electricity prices.

7. Argument: Steep reductions in greenhouse gases cannot occur without a significant increase in subsidies for nuclear power.

    Response: Nuclear power is probably the most expensive form of low-carbon power. Energy efficiency, wind power, solar photovoltaic, and concentrated solar thermal electric can achieve all the reductions that we need in the utility sector for the foreseeable future and at a lower price.

These arguments all have practical solutions, leaving industries and their responses visible and accountable to the public sector.

(DOE), eight coal-burning plants were canceled in October 2007 and eight others around the country were called off after May 2008.

Joyce Herms of Clean Wisconsin said, "It's not quite a trend, but a real indicator as to the direction things may be going."

Industry Ignoring Their Own Scientists

According to a *New York Times* article on April 24, 2009, by Andrew C. Revkin, the Global Climate Coalition—a group that represents industries tied to fossil fuels—has been involved in aggressive lobbying against the idea that the emissions of heat-trapping gases into the atmosphere have led to global warming.

In the 1990s, the group released public statements designed to confuse and add doubt to the issue in order to sway public opinion and lessen the impact on what research scientists were warning concerning global warming. Because the media attempted to give equal time to differing opinions, the Global Climate Coalition was able to gain ground. It was financed by large oil, coal, and the auto industry—those who had a vested interest in lessening the publics' attention on the negative effects of global warming.

Many environmentalists believe that these industries have understood the connection of their business activities to the rise in greenhouse gases and global warming all along, but that these industries treated the issue similarly to how tobacco companies treated the lung cancer issue years ago by insisting that "the science linking cigarette smoking to lung cancer was uncertain." In a similar way, the Global Climate Coalition has been able to cast enough doubt on the issue that a public outcry has been averted.

According to George Monbiot, a British environmental activist, "They didn't have to win the argument to succeed, only to cause as much confusion as possible."

In 2002, the coalition formally disbanded. Members, such as the American Petroleum Institute still individually lobby against global warming; but others, such as ExxonMobil, now recognize the human component of global warming and no longer financially support groups that challenge the science. It was discovered in a report prepared by the Global Climate Coalition in 1996 that greenhouse gas emissions-induced climate change was occurring and being affected by human

(continues on page 146)

THE OTHER GLOBAL WARMING—A LOOK INTO THE FUTURE

Even if the man-made greenhouse effect is brought under control, the Earth will still warm up, thanks to a completely different source of heat that humans create, according to Tufts University astrophysicist Eric J. Chaisson.

Chaisson predicts that over the next 250 years the Earth's population will start generating so much of its own heat—mainly from wasted energy use—that it will warm the Earth even without a rise in greenhouse gases. His advice is that the only way to avoid this is to rethink the way we use energy.

Chaisson chose to focus on waste heat that has not been studied in terms of global warming. He states that everything that uses or generates energy—such as cars, snow blowers, computers, lightbulbs, toasters, TVs, stereos, blow-dryers, iPods, and digital picture frames—squanders energy as wasted heat. He emphasizes that the larger and more energy-hungry the human population becomes, the more waste heat collects in the atmosphere.

According to Dennis Bushnell, the chief scientist at NASA's Langley Research Center, who supports Chaisson's ideas, said, "What this means for humans is that this is the ultimate limit to growth. As we produce more kilowatts, we have to produce more waste heat."

Chaisson's suggestion as a solution to the problem is not necessarily a cutback in energy use, but to change the way humans look at energy consumption. He believes that the population needs to shift toward power sources that do not add new heat to the Earth's system.

Chaisson contends that the waste heat problem is not all attributed to dirty fossil fuels such as coal, but also to some clean power sources such as nuclear and geothermal energy. Therefore, the only way to avoid adding extra unwanted heat is to use energy sources that already reach the Earth's surface. In other words, all of the energy used should originate only from sunlight and the wind and the waves that it powers.

Critics of Chaisson's theory say it is a scenario so far into the future that it is difficult to predict the extent to which it will play out. They also warn that attention should not be drawn away from present-day global warming at this point. Chaisson's idea, however, has captured the attention of many scientists worldwide. They see it as an opportunity to avoid a crisis before future generations have to face it. In a bigger sense, it also sets up a broader framework for decisions—one that looks at the environment from a long-term perspective.

Chaisson credits the basis for a lot of his ideas to his mentor and friend, the late astronomer Carl Sagan. He links his ideas to a fundamental law of science—energy cannot be perfectly harnessed, but tends to dissipate, usually in the form of heat. This concept, also called entropy, is dealt with in the second law of thermodynamics.

While the wasted heat today is not a huge problem, in the future it will be. The more energy humans use to fuel society and feed populations, the more waste heat will be emitted. As more countries industrialize and populations grow, the heat eventually will become a significant problem. Chaisson believes that even if all greenhouse gas pollution was capped, in 300 years the Earth will still be 5°F (3°C) warmer.

Mark Flanner, a research scientist at the National Center for Atmospheric Research in Boulder, Colorado, agrees with Chaisson's theory. Although he believes it is not a major problem right now, he supports the idea that waste heat will play a much larger role over the long term. He said, "If we continue the current growth rate in nonrenewable energy use, the heat flux will be of equal magnitude to the greenhouse effect 200 years from now."

Yangyang Liu, an atmospheric scientist at Brookhaven National Laboratory, who was also intrigued by Chaisson's theory, but thinks it may overestimate waste heat's contribution to long-term global warming, said, "We just don't know if there is a point at which energy use will level off. The efficiency to convert energy to work will also probably improve over time."

John Merrill, an atmospheric physicist at the University of Rhode Island's School of Oceanography, said, "I can't show that the estimates are wrong, but I can say that there are many hurdles caused by climate change and other environmental and social challenges that we need to address a lot sooner than this set."

(continues)

(continued)

Chaisson concedes that in looking 300 years into the future, his critique is outside current debates over climate change. But he warns that taking a long-term view is vital to human survival—even if the coming environmental catastrophe is something that neither we, nor our children, are likely to see.

(continued from page 143)

behavior and activities. The committee approved the final document, but only after the section acknowledging anthropogenic global warming was removed. According to the minutes of the meeting, "This idea was accepted and that portion of the paper will be dropped."

When questioned, William O'Keefe, the chairman of the Global Climate Coalition and a senior official of the American Petroleum Institute, said he was not aware that part of the final report had been deleted.

Benjamin D. Santer, a climate scientist at Lawrence Livermore National Laboratory, whose work for the IPCC was challenged by the Global Climate Coalition said, "I'm amazed and astonished that the Global Climate Coalition had in their possession scientific information that substantiated our cautious findings and then chose to suppress that information."

Clean Coal Technology

Coal is the major source of electricity in the United States and China. One solution that has been proposed to clean up the pollution from coal to combat global warming is clean coal technology. Coal is the dirtiest of the fossil fuels. According to the Energy Information Administration (EIA), coal generates half of the electricity in the United States and will likely do so as long as it is cheap and plentiful. According to the U.S. government, clean coal technology strives to reduce the harsh environmental effects by using multiple technologies to clean coal and contain its emissions. Not everyone agrees, however, that clean coal is actually clean. Many groups criticize the technology, not convinced that it can do what it claims it can.

Most coal—92 percent of the U.S. supply—is used in power production. Power plants burn coal to make the steam that turns turbines and generates electricity. When coal burns, it releases CO_2 and other emissions in flue gas. Some clean coal technologies claim to purify the coal before burning it. One method—called coal washing—removes unwanted minerals by mixing crushed coal with a liquid and letting the impurities separate from the coal and settle out to be removed. Other systems control the way the coal is burned in order to minimize the sulfur dioxide, nitrogen oxide, and particulate emissions. Wet scrubbers are also used. Also referred to as flue gas desulfurization systems, they remove sulfur dioxide—a major cause of acid rain—by spraying flue gas with limestone and water. The mixture reacts with the sulfur dioxide to form a synthetic gypsum (a component of drywall).

Low nitrogen oxide burners can also be used to reduce the creation of nitrogen oxides. Electrostatic precipitators are used to remove particulates that aggravate asthma and cause respiratory problems for people by charging particles with an electrical field and then capturing them on collection plates.

Gasification processes avoid burning coal altogether. With integrated gasification combined cycle (IGCC) systems, steam and hot pressurized air or oxygen combine with coal in a reaction that forces carbon molecules apart. The resulting syngas, a mixture of carbon monoxide and hydrogen, is then cleaned and burned in a gas turbine to make electricity. The heat energy from the gas turbine also powers a steam turbine. According to the U.S. Department of Energy (DOE), since IGCC power plants create two forms of energy, they have the potential to reach a fuel efficiency of 50 percent.

According to a report on CBS News, the cleanest coal plant in North America is operated in Florida by Tampa Electric. They call it clean coal, but they do not exactly burn coal. The process they use involves mixing coal with water and oxygen and converting it into a gas. John Ramil, Tampa Electric's president, said, "Gasifying coal allows the company to remove pollutants like sulfur, nitrogen, and soot, which virtually eliminates acid rain. And you can do it much cleaner than with the conventional coal technology."

James E. Hansen of GISS has another view on the issue. "There is no such thing as clean coal. All coal plants still emit millions of tons of CO_2—the most threatening greenhouse gas. There is no coal plant that captures the CO_2 and that's the major long-term pollutant."

Other scientists believe it is possible to recover most of the CO_2 and store it underground—CCS. In Norway, one company is storing CO_2 in rock caves beneath the North Sea. The DOE was planning a CCS project but halted it when the cost became too high.

In a study that appeared on the Web site LiveScience, a list was given of both the pros and cons of clean coal technology and related issues. The pros for pursuing clean coal technology are that the United States relies heavily on coal as a dependable source of power. Barbara Freese of the Union of Concerned Scientists (UCS) states that because of the United States' heavy reliance on coal as an energy source, "Weaning ourselves off of coal is obviously going to be very difficult and take some time. Given the unprecedented, urgent threat we face with climate change, we can't afford to ignore any technological option that could be part of the solution. We might need coal technologies to reduce our CO_2 emissions more quickly than we could if our only technological options were renewable energy sources and energy efficiency strategies."

The cons of clean coal technology point to the fact that coal is a dirty fuel and not easy to clean. According to Freese, "You're taking an inherently very polluting fuel, with each pollutant posing myriad problems and solving each with different technologies, and that keeps adding up in terms of cost.

"Even if CCS works properly with coal exhaust, you're depending on a nonrenewable resource for energy, and one that's notoriously destructive of the environment when it comes to mining it out."

Industry and cities have several hurdles to overcome. As new technologies become more advanced, implemented, and affordable, shifts need to be made to cleaner, sustainable, renewable energy sources that do not have a negative impact on the environment.

Global Warming: Agriculture Today

Agriculture plays a significant role in global warming, not only as a source of greenhouse gases (GHG) and pollution, but also as a source of renewable energy. This chapter examines the ways that agriculture affects and is affected by global warming. It also takes a look at which global warming issues are affecting the world today and how people getting involved and making a difference can help solve the problem.

AGRICULTURAL GREENHOUSE GASES AND POLLUTION

One of the most significant sources of methane from agriculture comes from stock manure. The decomposition of animal waste in an anaerobic (oxygen-free) environment produces methane. According to the U.S. Environmental Protection Agency (EPA), manure storage and treatment systems account for about 9 percent of total U.S. emissions and one-third of all methane emissions in the agricultural sector.

Liquid-based manure systems, such as manure ponds, anaerobic lagoons, and holding tanks, account for more than 80 percent of total methane emissions from animal wastes. Solid manure management practices (such as spreading manure across the surface of fields) produces insignificant amounts of gas, but it can lead to increased nutrient runoff, which can have a negative effect on water quality. From 1990 to 1996, emissions from manure management increased by 11 percent as farm animal populations grew and farmers expanded their use of liquid manure management.

An anaerobic digester is a container, like a covered lagoon. Methane produced by digesters, known as biogas, can be captured cost effectively and used as an energy source. Biomass recovery systems trap the gas in covered manure lagoons or other manure digesters, collect it in perforated pipes, and transmit it to an electric generator or boiler.

Farmers can use biogas to produce electricity, heat, hot water, and refrigeration for use on the farm, while at the same time controlling methane emissions and surface/groundwater contamination. Electricity can also be sold to utilities, and the digested solids—a high-quality fertilizer—can be sold to other farmers, home gardeners, or landscape designers.

The federal AgSTAR program, a joint initiative of the EPA, the U.S. Department of Energy (DOE), and the U.S. Department of Agriculture (USDA), teaches farmers how to manage manure profitably and protect the environment at the same time.

To make sure that methane recovery systems are designed, installed, maintained, and operated correctly, the USDA's Natural Resources Conservation Service and the EPA have developed conservation practice standards for methane recovery systems.

Globally, agriculture is responsible for 20 percent of greenhouse gas emissions. In the United States, the national average from agriculture is 8 percent. Agriculture emissions come primarily from methane and nitrous oxide. While these gases exist in smaller quantities in the atmosphere than CO_2, they are much more potent, which makes them serious greenhouse gas contenders.

A recent report from the United Nations indicates that livestock are responsible for 18 percent of greenhouse gas emissions worldwide. The

United Nations contends that it is also going to get much worse. As living standards climb in the developing world, the demand for meat and diary products increases as well. To back this claim, they report that annual per capita meat consumption in developing countries doubled from 31 pounds (14 kg) in 1980 to 62 pounds (28 kg) in 2002, based on data compiled by the Food and Agriculture Organization of the United Nations (FAO). The report predicts that global meat production will more than double by 2050. These statistics mean that the environmental damage from ranching would have to be cut in half just to keep emissions at their current, dangerous level.

The truth is that cattle, sheep, goats, and other ruminants naturally expel methane and nitrous oxide. It is estimated that a single cow can belch out anywhere from 25 to 130 gallons of methane a day. Farmers do have some ways of reducing this negative environmental impact. Improving their cattle's diet can improve methane and nitrous oxide emissions. Another option to reduce emissions is to capture and destroy methane that may have been created in manure lagoons. The greatest opportunities so far, however, are to use farms to produce *biofuels* and displace the emissions from fossil fuels used by other sectors.

RENEWABLE ENERGY GENERATION

Farmers are now being offered opportunities not only to produce on their land as they have always done, but now to enter other business ventures on their land by generating renewable energy. Wind power is one of the most rapidly growing renewable energy types in the world today. From 1998 to 2002, its annual growth rate worldwide was 32 percent. The DOE's Wind Powering America initiative has set a goal of producing 5 percent of the nation's electricity from wind by 2020. The DOE projects meant to achieve this goal will provide $60 billion in capital investment to the rural United States, $1.2 billion in new income to farmers and rural landowners, and 80,000 new jobs during the next 20 years.

Wind power can provide an important economic boost to farmers. Large wind turbines typically use less than half an acre of land, including access roads, so farmers can continue to plant crops and graze livestock right up to the base of the turbines.

Another form of clean energy that can be used on a farm is solar energy. According to the Union of Concerned Scientists (UCS), solar energy can be used in agriculture in a number of ways. One of the simplest ways is to design or renovate buildings and barns to use natural daylight instead of electric lights. Dairy operations using long day lighting to increase production can save money with skylights and other sunlighting options. Solar energy can also be used to warm homes and livestock buildings. Active solar heating systems, which use heat boxes and fans, can warm the air, saving fuel. Passive solar designs, where the building is situated to take advantage of the Sun automatically, are usually the most cost-effective approach.

The Sun can also be used to dry crops and grain, as well as heat greenhouses. Sunlight can also be used to generate electricity. Photovoltaic (PV) panels are often a cheaper option for providing electricity and are often used successfully for pond aeration, small irrigation systems, and remote livestock water supply.

Conversion technologies also exist now to convert perennial grasses into biofuels. According to the Pew Center on Global Climate Change, improvements in conversion technology could be achieved with additional research. They believe that dedicated energy crops could supply up to 12 percent of the current U.S. energy demand without significantly raising food prices. The biomass could be used to produce heat, power, and transportation fuels.

The largest use of biomass for energy today is cogeneration of steam and electricity by the forest products industry. Ethanol made from corn grain is the largest source of biomass-derived fuel in the transportation market. The Pew Center has calculated that crops could be grown specifically for use in energy production, with perennial grasses being the most commonly used vegetation source. They also support the idea that if aggressive research and development programs are put in place and move forward and successfully increase yields of bioenergy crops while reducing the current costs of biomass-to-fuel technologies, biomass from agricultural sources could supply as much as 19 percent of the total current U.S. demand for energy. This includes energy from dedicated energy crops, dual-purpose crops (such as corn and soybeans),

A combine harvesting corn to be converted into biofuels *(Warren Gretz, DOE/NREL)*

agricultural residues, and animal wastes. This amount of bioenergy represents an amount equivalent to more than 80 percent of the petroleum energy used for transportation in the United States.

The U.S. ethanol industry has grown from being almost nonexistent in the late 1970s to a current production level of more than 2.8 billion gallons—which equals about 2 percent of the energy consumed on America's highways each year. Ethanol made from corn grain is the largest source of biomass-derived fuel in the U.S. transportation market. Grain ethanol is made by converting the starch in the corn kernel to sugar and then using yeast to ferment the sugars to ethanol.

USDA and DOE researchers published a joint study of the potential magnitude of biomass supplies. They conservatively estimated the maximum economically feasible supply of ethanol from cornstarch at approximately 10 billion gallons (38 billion liters) per year. Some critics have argued that the process of creating ethanol from cornstarch requires more energy to make than it was worth, but according to the Pew Center, over the past 20 years, technology has advanced to the point that it provides more power than it takes to produce, making it a positive energy source.

Ethanol is also made from sugarcane. The energy balance for sugar-derived ethanol is even better than that for corn-based ethanol. It has a better energy balance because its bagasse (biomass that remains after the sugar has been extracted from the cane), rather than a fossil fuel, is used as the energy source in the conversion process. For corn-starched derived ethanol, fossil fuels are used in the conversion process.

A Hawaiian biomass gasifier that uses residue from the nearby sugar-cane mill as its energy source *(Rich Bain, DOE/NREL)*

Agricultural areas are now considered prime targets for producing four types of renewable energy: biodiesel, biomass, ethanol, and biogas. The diesel engine, created in 1892 by Rudolf Diesel, was originally designed to run on peanut oil. Today, diesel engines can be run on vegetable oil, even if it has already been used for cooking. Over the years, this has generated an interest in alternative fuels and the subsequent creation of biodiesel. Biodiesel is comprised of methyl esters of long-chain fatty acids and is derived from vegetable oil or animal fat by removing the glycerin in a reactive refinery process. Today, with little or no modification, diesel engines can run on a blend of petroleum diesel and biodiesel.

Biodiesel can also be burned in furnaces to heat homes, hothouses, or barns with only slight modifications to hose connectors and nozzles. The benefits of biodiesel include better air quality because of a reduction in particulates, volatile organic compounds, sulfur dioxide, carbon monoxide, and mercury. Biodiesel is also biodegradable and nontoxic. It also has increased lubricity and therefore reduces engine wear. The recent availability of biodiesel has improved, and prices are at least competitive, but often better, than regular diesel prices.

Biomass consists of a family of fuels whose plant-based feedstocks are either burned directly—such as wood or woody crops, corn, and chipped/palletized residue—or refined so that fuel in the form of sugar and starches are separated from the cellulose.

Cellulose—the primary component in biomass fuels—is found in all plants. When processed, it is the cellulose that provides the energy for heating, transportation, and electrical generation. Many types of materials can be converted to biomass, such as wood, wood crops, waste wood, wood residue, grains, grasses, and agricultural residue.

Biorefineries for biomass are currently being researched as many scientists are focusing on them, seeing their importance long into the future. In the future, scientists expect to be able to extract from plants chemicals and fuels through the depolymerization of the cellulose and hemicellulose and the conversion of lignin to fuel and synfuels. While the technology is not ready yet, the pressure to develop this, in light of global warming and the need for greenhouse gas reductions, as well as opportunities for a new farm economy, should lead to rapid development in this area.

Napier grass (also called elephant grass) is used for the production of ethanol. *(Warren Gretz, DOE/NREL)*

Ethanol, like biodiesel, is a biofuel that can serve as an alternative to petroleum fuels, although it can only be added to or substituted for gasoline, not diesel. Most of today's ethanol is produced from starch-based feedstock such as corn or sugarcane. It can also be produced from cellulosic feedstock such as fast-growing grasses like the perennial switchgrass or short-rotation woody crops like poplar.

Biogas is a gas produced by the biological breakdown of organic matter in anaerobic conditions. It originates from biogenic material and is a type of biofuel. One type is produced by the anaerobic digestion or fermentation of biodegradable materials such as sewage, manure, municipal waste, or energy crops. The gas is composed mostly of carbon dioxide (CO_2) and methane. Another principal type of biogas is called wood gas, which is created by gasification of wood or other biomass. This biogas is composed of nitrogen, hydrogen, and carbon monoxide, with small amounts of methane.

Biogas is commonly used for heating, to run heat engines, and to generate mechanical or electrical power. It can also be compressed like natural gas and used as a fuel in cars. Biogas is also referred to as swamp, marsh, or landfill gas.

IN THE NEWS—WHAT IS HAPPENING TODAY

Several cities and states across the United States, as well as countries worldwide, are taking measures to fight global warming and achieving noteworthy results. Some examples follow.

- Texas has added more than 4,000 megawatts of wind power–generating capacity in the past 10 years. Wind power now provides 3 percent of Texas's electricity, which is enough to keep about 8.8 million tons (8 million metric tons) of GHG out of the atmosphere each year.
- New Jersey has doubled its solar power–generating capacity within the past two years through public policies that promote solar panels on rooftops.
- California uses 20 percent less energy per capita than it did in 1973, thanks to strong energy-efficiency policies for buildings and appliances.
- Wisconsin has adopted several environmental policies to promote energy efficiencies in industry. These programs have not only been able to save businesses money and have created new jobs within the state, but they have kept 220,462 tons (200,000 metric tons) of CO_2 out of the atmosphere.
- Portland, Oregon, has doubled the number of bicyclists in the past six years by making the city bicycle friendly.

- Improvements to the mass transit systems in Rosslyn and Ballston, Virginia, have encouraged about 40 percent of the residents to take mass transit.
- Southeastern Pennsylvania has a 20 percent increase in the number of passengers on the trains that travel to Harrisburg and Philadelphia because the travel speeds were increased, making the trains more efficient, reliable, and attractive to use.
- Germany has recycled 60 percent of its municipal waste for the past 20 years. They have enacted policies that put the responsibility on product manufacturers instead of individual consumers and taxpayers.
- In Israel, more than 90 percent of the homes use solar water heaters, which have greatly reduced the need for natural gas and electricity for water heating. Israel requires that all new homes be equipped with solar water heaters.
- In Copenhagen, Denmark, pedestrians and bicyclists have preference over cars in its downtown city center section. Currently, about 40 percent of the population walks or rides bicycles.
- Spain is now third in the world for wind farms and wind power capacity and is the world's fourth-leading market for solar PVs.

On November 1–2, 2007, America's mayors gathered in Seattle, Washington, for a summit to spur action on climate change. As of May 2009, 720 mayors across the United States have signed the US Mayors Climate Protection Agreement. Their influence has inspired assertive action by other institutions and organizations across the country. Some of the active participants include the Sierra Club's Cool Cities campaign and the American College & University Presidents Climate Commitment.

The United States Conference of Mayors has also established the Mayors Climate Protection Center to strengthen the agreement. The purpose of the center is to recruit additional participants, improve technical assistance to participating cities, develop a tracking and reporting system, and increase mayors' participation in federal policy making.

Seattle mayor Greg Nickels created the Seattle Climate Action Plan in September 2006 and allocated $41 million in investments for 2007 and 2008, including $38 million from a voter-approved transportation funding package. The main focus of the action plan is to reduce communitywide greenhouse gas emissions by at least 7 percent below 1990 levels by 2012. Some of the initiatives in the action plan include the construction of green buildings, the creation of bike- and pedestrian-friendly neighborhoods, the enhancement of public transportation, the creation of city recycling programs, and the initiation of water conservation measures. Currently, the Seattle city government has announced that Seattle has reduced its global warming pollution by about 60 percent compared to 1990 levels.

From the perspective of CO_2 levels, a lot has been in the news lately. According to an article on the Web site ScienceDaily, from 2007, atmospheric CO_2 growth has increased 35 percent faster than expected since 2000. On April 24, 2008, NOAA posted a news release calling for better monitoring of CO_2 around the world. With global CO_2 concentrations now at 385 parts per million (ppm) and rising, effective CO_2-monitoring strategies are almost nonexistent.

Preindustrial CO_2 levels were approximately 280 ppm until 1850. Human activities pushed those levels up to 380 ppm by early 2006. The rate of CO_2 increase has accelerated over recent decades because of fossil fuel emissions. Since 2000, annual increases of two ppm or more have been common, compared with 1.5 ppm per year in the 1980s and less than one ppm per year during the 1960s. This rate is expected to increase with growing industrialization in India and China. In fact, the significant jump in CO_2 levels from 2006 to 2007 is attributed to Chinese emissions, which account for two-thirds of the increase. In 2006, China surpassed the United States to become the biggest emitter of greenhouse gases—roughly 24 percent of the total.

One of the major industries responsible for this in China is the cement industry, and China accounts for about 51 percent of global cement production. In 2007, China emitted 2 billion tons (1.8 billion metric tons) of carbon from fossil fuels, compared with 1.59 billion by the United States. If China's carbon use keeps up with its booming economic growth, the country's CO_2 emissions will reach 8 gigatons a year by 2030, which is what the entire global CO_2 production is today. China is expanding so fast they are adding one to two new coal-fired power plants to their grid each week. China's emissions increased 8 percent in 2007. A report released by the Netherlands Environmental Assessment Agency found that in 2007 China's emissions were 14 percent higher than those of the United States.

The power being generated is being used to drive an enormous manufacturing system. For example, it has increased steel production from 140 million tons (127 million metric tons) in 2000 to 419 million tons (380 million metric tons) in 2006. In 1999, the Chinese bought 1.2 million cars. In 2006, they bought 7.2 million cars.

China's industrial growth is already having ill effects. In Handan City, 300 miles (483 km) south of Beijing, one of the biggest steel mills in China is operated. Handan is a highly populated city. The residents that reside on the west side, downwind of the steel mill, are caught in a steady flow of dust and smoke full of carcinogens. After repeated public outcry over the unhealthy living conditions, the mining company agreed to compensate the affected residents with a pollution fee rather than clean up the source of the pollution. According to Tian Lanxiu, a resident of Handan, "The villagers have learned to cope with Handan's emissions. People do not eat outdoors to avoid having black briquettes flake their rice." She also says that if her children cannot sleep at night, she stuffs their ears with cotton. She says many people in the village die young—often of heart disease or cancer. Although she has no evidence to connect their deaths to the steel mill, she has few doubts herself. "Handan knocks 10 years off people's lives. We all want to live longer. We're growing more aware."

According to a report from the DOE's Oak Ridge National Laboratory (ORNL), annual CO_2 emissions from burning fossil fuels and manufacturing cement have grown 38 percent since 1992, from 6.1 billion tons (5.5 billion metric tons) of carbon to 8.5 billion tons (7.7 billion metric tons) in 2007.

Gregg Marland at ORNL's environmental sciences division reported on September 24, 2008: "The United States was the largest emitter of CO_2 in 1992, followed in order by China, Russia, Japan, and India. The most recent estimates suggest that India passed Japan in 2002, China became the largest emitter in 2006, and India is poised to pass Russia to become the third largest emitter, probably this year." ORNL has determined that the latest estimates of annual emissions of CO_2 to the atmosphere indicate that emissions are continuing to grow rapidly and that the pattern of emissions has changed markedly since the drafting of the United Nations Framework Convention on Climate Change (UNFCCC) in 1992. It was then that the international community expressed concern about limiting emissions of greenhouse gases.

As the CO_2 emissions continue to increase, the source of these emissions has shifted dramatically as energy use has skyrocketed in developing countries—most notably in the rapidly developing Asian countries. This is one of the principal reasons why George W. Bush did not ratify the Kyoto Protocol—it excluded both China and India from having to be held accountable for their CO_2 emissions; now China is the largest CO_2 emitter on Earth.

The United Nations reported that a huge, thick brown cloud of soot, particles, and chemicals stretches from the Persian Gulf to Asia, threatening the health of the people living there. Cities such as Cairo, New Delhi, Mumbai, Beijing, and Shanghai have their light dimmed by as much as 25 percent.

The World Energy Outlook 2008 from the International Energy Agency (IEA) has predicted that China and India will account for just over half of the increase in world primary energy demand between 2006 and 2030. Middle Eastern countries will account for an 11 percent incremental increase in world energy demand. If China continues to increase their CO_2 emissions through the burning of dirty coal, at 10 percent per year growth, they could double their emissions within the next seven years.

On April 2, 2007, in one of its most important environmental decisions in years, the U.S. Supreme Court ruled that the EPA has the authority to regulate heat-trapping gases in automobile emissions. The Supreme Court carried it a step further in giving responsibility to the EPA. They also ruled that the agency could not "sidestep its authority to regulate the greenhouse gases that contribute to global climate change unless it could provide a scientific basis for its refusal."

Justice John Paul Stevens said the only way the agency could "avoid taking further action" now was "if it determines that greenhouse gases do not contribute to climate change" or provides a good explanation why it cannot or will not find out whether they do. The decision is expected to have a broad impact on the debate over whether the federal government needs to directly address the issue of global warming or not. To date, court cases nationwide have been put on hold waiting to see what the EPA will do. One case challenges the EPA's refusal to regulate CO_2 emissions from power plants, which is now pending in the federal appeals court. Justice Stevens states that if the EPA only provides a "laundry list of reasons not to regulate," the EPA is defying the Clean Air Act's "clear statutory command." He said a "refusal to regulate could be based only on science and reasoned justification," and that while the statute left the central determination to the "judgment" of the agency's administrator, "the use of the word 'judgment' is not a roving license to ignore the statutory text."

On May 19, 2009, President Barack Obama proposed the most aggressive increase in United States auto fuel efficiency ever in a policy initiative that regulates emissions for the first time, as well as resolving a dispute with California over cleaner cars. The initiative will require that cars and light trucks increase efficiency by 10 miles (16 km) per gallon to 35.5 miles (56.8 km) per gallon between 2012 and 2016.

With these standards in place, global warming carbon emissions are expected to decrease by 992 million tons (900 million metric tons). The program does not just target one company—it involves all car manufacturers. All companies will be required to make more efficient and cleaner cars, which will equate to a savings of 1.8 billion barrels of oil.

According to Senator Barbara Boxer, chairman of the Environment and Public Works Committee, this new development is "good news for all of us who have fought long and hard to reduce global warming and reduce U.S. dependence on imported oil."

The Obama administration opened the way to regulating emissions by declaring climate-warming pollution a danger to human health and welfare.

David Friedman, research director of the clean vehicle program at the Union of Concerned Scientists (UCS), said, "This could be the breakthrough we've been looking for on clean cars."

Automakers are actively pursuing the manufacture of better hybrids and electric cars. One recent concern is that the decline of gas prices toward the end of 2008 has encouraged consumers to buy less-efficient trucks and SUVs again. In addition, the downward turn in the economy has led to lower sales of the slightly more costly hybrid vehicles. To help lift the industry out of it current slump, the federal government offered the Cash for Clunkers program that ran from July 1 through August 24, 2009. The program offered $3,500 or $4,500 to those who qualified to trade in their old, low-mileage car for a new, fuel-efficient car. The program was very successful—nearly 700,000 less fuel-efficient cars were traded in and hence taken off the nation's highways.

Another disturbing event reported in the *New York Times* is that several European countries have announced plans to divert back to coal-fired plants regardless of coal's reputation for being the "dirtiest fuel on Earth." Italy's major electricity producer, Enel, is now in the process of converting its power plant to run on coal instead of oil. They claim the reasons to be the skyrocketing prices of oil, the lack of energy security in having to deal with sometimes-hostile foreign countries to obtain oil, and their distrust and aversion to nuclear power as an energy option. Therefore, over the course of the next five years, Italy plans on increasing its reliance on coal from its present level of 14 percent to 33 percent.

Unfortunately, Italy is not the only European country with these plans. Other countries are planning to put into operation another 50 coal-fired plants over the next five years for the very same reasons Italy is. These plants will remain in production for the next 50 years.

The scene in the United States is somewhat different. Fewer coal plants are likely to be constructed in the future for two primary reasons: It is getting more difficult to get regulatory permits, and nuclear power is still an alternative power source. Of the 151 proposals for coal-fired plant constructions across the United States in early 2007, more than 60 had been dropped by early 2008, the bulk of them stopped by state governments. Dozens of others are frozen in legal battles.

The situation in Europe is causing an uproar among environmentalists, prompting several protests. There is even a new plant being constructed in Kent—which is planned to become the United Kingdom's first new coal-fired plant in more than a decade. The power station promoters in Europe who are behind the plants stress that they are building the new coal plants as "clean as possible." Critics, however, are not buying that and are saying, "Clean coal is nothing but a pipe dream—an oxymoron in terms of the carbon emissions that count most toward climate change." They call this sudden building spurt "extremely shortsighted."

James E. Hansen had this to say: "Building new coal-fired power plants is ill conceived. Given our knowledge about what needs to be done to stabilize climate, this plan is like barging into a war without having a plan for how it should be conducted, even though information is available. We need a moratorium on coal now, with phaseout of existing plants over the next two decades."

According to the Electric Power Research Institute (EPRI), under optimal current conditions, coal produces more than twice as much CO_2 per unit of electricity as natural gas, the second most common fuel used for electricity generation. In the developing world, where even new coal plants use low-grade coal and less efficient machinery, the equation is even worse. Simply put—without carbon capture and storage (CCS), coal cannot be green.

According to Jeffrey D. Sachs, director of the Earth Institute at Columbia University, "Figuring out carbon capture is really critical—it may not work in the end—and if it is not viable, the situation with respect to climate change is far more dire."

There are a few demonstration projects in the United States and Europe. At the end of January 2008, the Bush administration cancelled what was by far the United State's biggest carbon capture demonstration project—a coal-fired plant in Illinois—because of huge cost overruns. The cost of the project, begun in 2003 with a budget of $950 million, had spiraled upward to $1.5 billion this year, and it was far from complete.

Many experts believe CCS technology is a challenge similar to putting an astronaut on the Moon. Norway, which is currently investing heavily to test the technology, calls carbon capture its "moon landing." Others, however, think it may be even harder than that—it is a moon landing that must be replicated daily at thousands of coal plants in hundreds of countries. Another problem is that CCS is expensive, and some countries, such as China and India, may not be able to afford the technology, even if it were developed and available. According to the EPA, plants that could someday be adapted to carbon capture would cost 10 to 20 percent more to build and only a handful exist today. For the majority of coal plants, the cost to convert would be phenomenal. Finding appropriate, viable storage sites that would safely contain the carbon without leaking would also be a challenge. The site would have to be large enough to store the waste and not be subject to movements within the Earth or weakening of its crust.

MAKING A DIFFERENCE: SOLUTIONS TO THE PROBLEM

There is an overwhelming consensus in the scientific community that global warming needs to be addressed immediately. As technology continues to improve, there is no doubt that it will be able to help make transitions to a greener lifestyle easier. But it is not necessary to wait for technology to solve the problems of global warming. Because of projections that CO_2 emissions must be cut in half in the next few decades, action must be taken now.

Two scientists from Princeton University—Stephen Pacala and Robert Socolow—have come up with a plausible solution to the problem, geared to involving the world's population now. The visual representation of their solution is called the stabilization wedges graph and can be seen on the next page.

In the graph, every wedge represents a global warming solution that is geared toward stabilizing heat-trapping emissions at a safe level. The idea represented here is that if the immense problem of global warming is tackled in small enough individual chunks, the individual sums of all the changes will add up significantly. Their graph illustrates that by employing increased efficiency, using more renewable energy, capturing and storing CO_2, and other measures, the Earth's population can collectively reduce CO_2 levels to a safe concentration.

The Intergovernmental Panel on Climate Change's (IPCC) 2007 report laid out various guidelines toward the solution of the global warming problem. Various solutions consist of making changes in what kind of energy is produced, how efficiently it is used, how often personal vehicles should be driven, how homes should be made energy efficient, and how humans will use the land and its natural resources in the future. If correct choices are made along these lines, then the

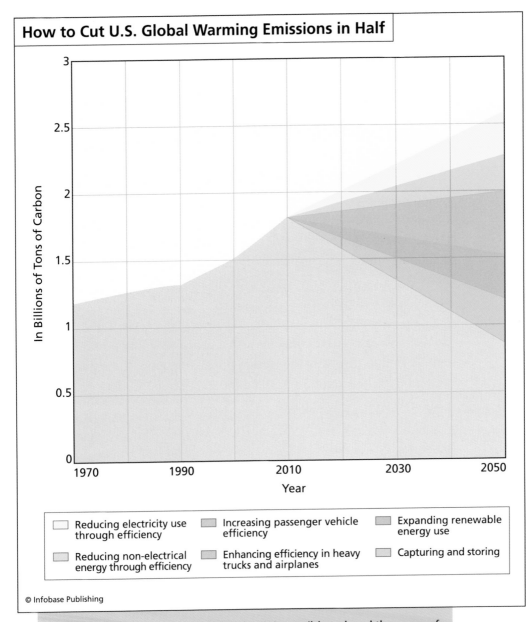

How to Cut U.S. Global Warming Emissions in Half

Legend:
- Reducing electricity use through efficiency
- Reducing non-electrical energy through efficiency
- Increasing passenger vehicle efficiency
- Enhancing efficiency in heavy trucks and airplanes
- Expanding renewable energy use
- Capturing and storing

© Infobase Publishing

Based on Pacala and Socolow's theories, it is possible to bend the curve of climbing global warming emissions downward by employing increased efficiency, more renewable energy, and other solutions. *(Source: NRDC)*

stabilization wedges will grow over time, effectively reducing the CO_2 concentration in the atmosphere.

In terms of power generation, Pacala and Socolow point out that with the long lifetimes of power plants, investment decisions will have enormous long-term impacts on the amount of global warming pollution allowed into the atmosphere. They support the use of clean, renewable energy. Fortunately, wind power, solar, biomass, and geothermal are among the fastest-growing sources of electricity globally, even though right now they only account for 2 percent of the electricity supply in the United States. Pacala and Socolow also recommend the development of technology that allows the carbon pollution from coal plants to be satisfactorily captured before it leaves the smokestacks.

Travel in personal cars must be made more efficient. By 2025, raising fuel economy to 35 miles per gallon (MPG) would reduce oil consumption by 2.3 million barrels per day, save consumers more than $56 billion per year, and reduce global warming pollution by the equivalent of 50 million vehicles. The Obama administration's new, aggressive automobile fuel standards for passenger cars and light trucks (including the popular SUV) are expected to be raised to 35.5 miles per gallon between 2012 and 2016. Incorporating more biofuels into the transportation sector would also help. Governments need to be involved—national, state, county, city, and community. Official planning must be put in place. Enhancing public transportation and creating walking-bicycling areas are also steps in the right direction.

Building design and efficiency are also important considerations. Researchers suggest new buildings use a building efficiency rating system as a guide, such as Leadership in Energy and Environmental Design (LEED) certification. It is calculated that the average building can cut electricity use by 32 percent and save 386 tons (350 metric tons) of carbon emissions each year. Installing more efficient lighting, electrical appliances, heating and cooling devices, and improved insulation will all result in reduced energy bills.

Using the Earth's open land areas properly can also add to the health of the planet. Conservation farming practices lead to less decomposition of organic surface matter, which allows more carbon to be stored in the soil. This has the potential to help reduce global warming. Keeping existing forests intact is another way to help fight global warming. As long as trees are left standing, they act as carbon sinks, storing carbon within them.

There are many environmental organizations involved in the battle against global warming. Several groups including American Rivers, National Audubon Society, Clean Water Action, Defenders of Wildlife, Environment America, Environmental Defense Fund, Friends of the Earth, Greenpeace, League of Conservation Voters, National Parks Conservation Association, Natural Resources Defense Council, National Wildlife Federation, Rails-to-Trails Conservancy, Sierra Club, Trust for Public Land, Union of Concerned Scientists, and the Wilderness Society share the belief that by protecting the environment it is possible to jump-start the economy, because it would create new jobs in clean energy and green infrastructure.

These organizations issued a green stimulus on December 11, 2008, in Washington, D.C., that would create up to 4 million new jobs, reduce pollution, save energy, protect public health and safety, and restore the environment.

The announcement was a $160 billion proposal for funding nearly 80 energy efficiency, renewable energy, public transportation, national parks and public lands, education, and other environmental programs for President Obama's transition team.

Shortly after Obama won the 2008 election, he said he wanted an economic recovery program, "building wind farms and solar panels, fuel-efficient cars and the alternative energy technologies that can free us from our dependence on foreign oil and keep our economy competitive in the years ahead."

The proposal offered by the conservation organizations is geared to do just that. They believe that "America's economic recovery hinges on making smart investments today that will protect our environment, restore our valuable natural resources, and lead us on the path to a clean energy economy." Rebecca Wodder of American Rivers said, "President Obama has the opportunity to not only create good jobs by making critically needed investments in energy, transportation, water, and lands, but also to lay the foundation for a new 21st century economy. We need smart investments to jump-start the economy now and to keep our country competitive in the years ahead."

"President Obama recognizes that with the challenge of rebuilding the economy comes the opportunity to repower America with clean, homegrown energy like wind and solar," said Anna Aurilio of Environment America. "Investing now in clean energy, energy efficiency, and smarter transportation is key to improving our energy security, solving global warming, and jump-starting our economy."

"Efficiency is the quickest, cheapest, cleanest way to reduce global warming pollution and has vast potential to create jobs and stimulate the economy," said Jim Presswood of the Natural Resources Defense Council.

Approximately 30 percent of U.S. global warming pollution and 60 percent of U.S. oil consumption are due to transportation. Rather than spending money on new roads, the groups recommend investing in green transportation projects and addressing America's billion-dollar backlog in road and bridge maintenance. Maintenance and rehabilitation create more jobs than new road construction and do so without increasing oil consumption and global warming emissions. "Investing in public transportation and other transportation alternatives, the next generation of alternative fuels, and vehicle efficiency can reduce our dependence on oil, reduce global warming pollution, save families money at the pump, and create millions of good jobs," said Colin Peppard of Friends of the Earth. "And when it comes to roads and bridges, the focus should be on maintaining existing infrastructure, not new construction."

"Climate change is already causing serious water shortages, flooding, and damage to ecosystems. We need to invest more in water infrastructure, but we need to invest more wisely, too. Building new dams isn't the answer. Instead, we must invest in green solutions like water efficiency and natural flood protection to create good jobs, save money, and protect public health and safety," said Betsy Otto of American Rivers.

The time to look toward the future is now. Decisions we make, systems we create, facilities we build, and laws we enact will affect not only us but also the generations to come. It is critical that society makes sustainable, intelligent choices now. While scientists are engaged in cutting-edge research, the time to boldly move ahead to help the populations of tomorrow is today.

Fuel Technology

As global warming grows as a threat, air and water pollution worsen, and energy costs continue to rise, scientists are diligently working on finding solutions to combat the problem. A major issue facing the United States is the current overreliance on petroleum-based transportation fuels. This chapter looks at alternative and advanced fuels such as biodiesel, ethanol, methanol, hydrogen, natural gas, and propane. It then examines where current research is headed with new fuel technology. Next, it focuses on what green technology is doing for the auto industry and explains and compares the differences between hybrids, flex-fuels, fuel cells, plug-ins, electrics, air-powered vehicles, and trendy future cars.

BIOFUELS AND CLEAN VEHICLES

It is well understood that one of the biggest contributors to global warming is the burning of fossil fuels and that the United States is one of the largest contributors to the problem. Transportation-related emissions are responsible for 40 percent of the U.S. total global warming pollution. Because of this large contribution, this is one area where people can make a significant difference. One way is by conserving and driving less, combining multiple errands into one trip, and by using new technology—more efficient vehicles and lower carbon fuels (fuels that generate far less heat-trapping gases per unit of energy). Hydrogen, electricity, and biofuels all have this capability. This also helps national security by reducing the country's dependence on foreign oil.

There is a wide range of characteristics among each alternative fuel type and their unique environmental emissions, if any, and their environmental impacts. Standards are currently being developed that will require fuel providers to account for and reduce the heat-trapping emissions associated with both the production and use of fuel.

California, the nation's largest market for transportation fuel, is developing a low carbon fuel standard that will require fuel providers to verify there is a reduction in global warming emissions per unit of energy delivered. When carbon emissions are calculated, all emissions are accounted for during the fuels' entire life cycles. The accounting system is very specific and also addresses any uncertainties. It also allows for changes over time as technology improves for assessments, as well as products, to become more refined. Through being able to keep track of the exact performance of each type of new fuel, it allows researchers to be able to assess just how effective the alternative fuels are toward making headway against global warming.

Smart fuel policies, such as California's, set to take effect in 2010, are important because they promote carbon reduction through the entire fuel-manufacturing process. According to the Union of Concerned Scientists (UCS), low-carbon fuel standards (LCFS) also create market certainty for cleaner fuels, making sure the fuel industry does its part along with the automakers and consumers—to reduce transportation emissions that relate to global warming. Other states that are also considering developing low-carbon fuel standards are Arizona, Minnesota, New Mexico, Oregon, and Washington.

LCFS are designed to work in tandem with the Obama administration's new auto fuel standards. Concerning the new fuel standards,

California supports them and has agreed that they will defer to the proposed national standard. In the new standard, tailpipe emissions will be cut by more than 30 percent, which is even greater than what California and other states had initially sought. What makes LCFS unique is that it deals with the entire life-cycle emissions of fuels on an average per-gallon basis. Instead of dictating specific technologies or fuel types, it allows suppliers lateral freedom to decide which methods they will use to meet the reduced emissions targets. The key aspect is the life cycle component. This means that every aspect involved in the fuel's retrieval, creation, and use must be accounted for in the way it contributes to global warming; not just one phase of the cycle, such as an automobile converting the gasoline into motion. In order to properly look at the true emissions of a fuel and its true carbon footprint, the fuel must be held accountable for (1) the emissions generated at the extraction source; (2) the refinery process; and (3) all the way to the tailpipe itself—a process dubbed from well-to-wheels.

In the case of biofuels, where the fuel is grown as an agricultural crop instead of mined from the ground, it must account for all of the emissions from tractors and fertilizers used to grow the crop, all the energy used to convert the crop to a fuel, and any other indirect sources of pollution—including any emissions generated from changes in land use as a result of biofuel production.

Because of all the interrelationships, accounting for all these inputs can become very complex, with many issues to deal with. However, the UCS stresses that these issues must be openly discussed and a consensus reached without any political lobbying or interference, to keep the technical issues and results reliable in order to avoid a distorted analysis. Otherwise, the desired goals of a biased study will keep the LCFS from being able to deliver the needed low-carbon fuels.

Under the LCFS program, fuel suppliers are not mandated to try particular technologies or specific fuels; fuel suppliers are free to choose how they meet their emission targets. For instance, they can choose to blend lower-carbon biofuels into gasoline to lower the carbon content; they can choose to reduce emissions from the refining process; or they can sell natural gas (which has a lower carbon content) for use as a transportation fuel. They also have the option of using trading credits, which provide even more flexibility and lower the cost of compliance if other methods, such as switching technologies, prove to be too costly. In this case, for example, fuel suppliers could purchase credits from electric utilities that supply low carbon electricity to plug-in hybrids.

An important aspect in the LCFS concept is that an accurate life cycle accounting must be kept, including any indirect emissions. Indirect emissions include situations such as when land must be cleared to grow crops that will be used for biofuels. If forested areas are removed, the lost carbon storage must be accounted for. If all the direct and indirect impacts to net carbon balance are not accurately accounted for, then calculations of carbon reductions may not be accurate. Otherwise, a study may declare a decrease in fuel emissions, yet actually contribute to an increase in global warming pollution.

The UCS believes that LCFS is important for three major reasons: (1) it promotes improvements in the supply chain; (2) it protects against high-carbon fuels; and (3) it creates choices and spurs innovation. When the full life cycle is considered, it offers the fuel provider opportunities to improve and lower the carbon content anywhere along the supply chain. It protects against high-carbon fuels by offering a clear incentive to use clean fuels over polluting fuels. For example, the coal-to-liquids technology has a life cycle global warming pollution almost double that of petroleum. It is a bigger incentive to choose a low-carbon fuel to begin with. This concept makes the dirtier fuels pay the price for their higher pollution. Because LCFS does not mandate using a specific approach or a certain technology, it opens the playing field and focuses only on companies' abilities to deliver cost-effective low-carbon fuel. This way it is not encumbered by government mandates that could slow new developments. Its design, instead, is to focus on who can supply the lowest-carbon fuels. According to the UCS, the investors and marketplace (public) will be the ones who will decide on the ultimate winners.

President Obama has also called for a nationwide low-carbon fuel standard to help meet his goal of cutting greenhouse gas emissions more than 80 percent by mid-century. California's LCFS would require refineries, producers, and importers of motor fuels sold in California to reduce the carbon intensity of their products 10 percent by 2020, with greater cuts thereafter.

At the national level, work is being done to encourage the information of heat-trapping emission requirements into the current renewable fuel standard. Several bills have also been introduced in Congress that would establish low-carbon fuel standards. Advanced clean vehicle technologies are available now. Hybrid-electric vehicles are available through Toyota and Honda. Fully electric vehicles are also available but are not as focused on because they are still rather expensive and have a limited driving range.

Fuel cells have come a long way in research. Various cities in the world have demonstrated fuel-cell bus programs, such as Chicago, Illinois, and Vancouver in Canada. In California, a partnership among automakers, the government, and fuel cell manufacturers is testing fuel cell technology and is expected to produce over 60 demonstration vehicles in the next few years.

In addition, the California zero emission vehicle (ZEV) program will require auto manufacturers to sell increasing numbers of zero emission vehicles over the next decade to further promote fuel cell vehicles. The trend has been set, and the public is voicing their eagerness to have these vehicles available. The majority of automobile manufacturers have announced their plans to begin selling fuel-cell passenger vehicles in the next few years.

There are two different concepts when referring to advanced clean vehicles—a fuel-efficient car and a low-emitting car. A higher fuel efficiency results in less global warming pollution. A low-emitting vehicle releases fewer smog-forming pollutants. The amount of fuel that a car burns determines how much carbon dioxide (CO_2) it releases. Air pollution–control devices on cars reduce other pollutants, such as carbon monoxide, or smog-forming pollutants, such as nitrogen oxides and volatile organic compounds (VOCs). All vehicles with high fuel efficiencies do not necessarily reduce urban smog—such as diesels. The vehicle must also be one that has low emissions. Truly green cars address all the issues: global warming, air pollution, and America's dependence on oil.

ALTERNATIVE AND ADVANCED FUELS

According to the U.S. Department of Energy (DOE), there are more than a dozen alternative and advanced fuels in production or being used today. As the general population becomes more educated and aware of their existence and availability and as the price of gasoline at the pumps skyrockets, there is a growing interest in using them as the green revolution gains momentum. These fuels provide numerous benefits—they are environmentally friendly and they lessen America's dependence on foreign oil sources. Alternative fuels have been defined by the Energy Policy Act (EPAct) of 1992. Alternative fuels that are commercially available for vehicles include biodiesel, methanol, propane, electricity, hydrogen, ethanol, and natural gas

The EPAct was passed by Congress to reduce the nation's dependence on imported petroleum by requiring specific vehicle fleets to

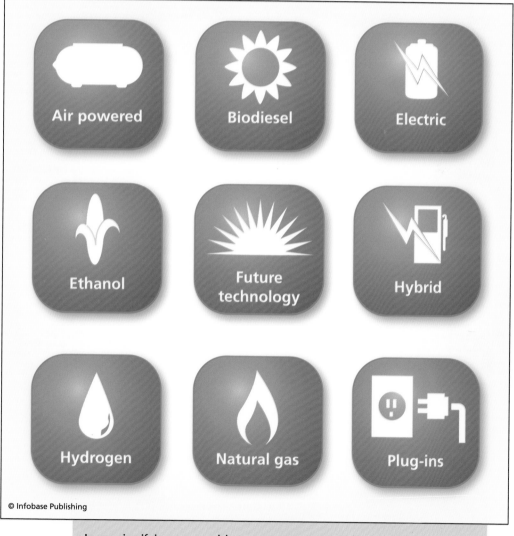

Icons signifying renewable energy *(Source: GreenCar.com)*

acquire alternative fuel vehicles, which are capable of operating on non-petroleum fuels.

Biodiesel

Biodiesel is a renewable alternative fuel that is made from soybeans, biomass, vegetable oil, animal fats, and recycled restaurant greases. Biodiesel can be used in its pure state (B100) or it can be blended with petroleum diesel. B2 is 2 percent biodiesel; there is also B5 and B20 blend. The liquid fuel is comprised of fatty acid methyl esters (FAME) or long-chain alkyl esters. Biodiesel is produced from renewable sources such as new and used vegetable oils and animals fats. It is also a much cleaner burning fuel than the traditional petroleum diesel, although it has physical properties similar to those of traditional diesel. Biodiesel is also nontoxic and biodegradable. Biodiesel can be used in most diesel cars without modification, making it an attractive choice of alternative fuel. The DOE has identified the following advantages and disadvantages of biodiesel.

Advantages and Disadvantages of Biodiesel	
ADVANTAGES	DISADVANTAGES
Domestically produced from non-petroleum renewable resources	Use of blends above 85 not yet warranted by automakers
Can be used in most diesel engines, especially newer ones	Lower fuel economy and power (10 percent lower for B100, 2 percent for B20)
Less air pollutants (other than nitrogen oxides) and greenhouse gases	Currently more expensive
Biodegradable	More nitrogen oxide emissions
Nontoxic	B100 generally not suitable for use in low temperatures
Safer to handle	Concerns about B100's impact on engine durability

The U.S. biodiesel industry is still relatively small but growing rapidly. From 2004 to 2005, production rates tripled, then tripled again from 2005 to 2006. By 2007, production had reached 491 million gallons (1.9 billion liters), which doubled the 2006 level. The bulk of biodiesel manufacture comes from industries involved in making products from vegetable oil or animal fat. One example is the detergent industry. The soy industry has been one of the major driving forces behind biodiesel commercialization because of overproduction of soy oils and falling prices in the market the past few years. According to the DOE, there is enough virgin soy oil, recycled restaurant grease, and other acceptable feedstock available in the United States to provide quality feedstock for approximately 1.7 billion gallons (6.4 billion liters) of biodiesel per year—which represents about 5 percent of the diesel used in the United States.

Electricity

Electricity can be used both to power electric and plug-in hybrid electric vehicles directly from the power grid. An important aspect of electric vehicles is that they do not produce any tailpipe emissions. The only emissions that can be attributed to electricity are those generated in the production process at the power plant that generates the electricity. The electricity option works well for short-range driving.

Electricity is used as the transportation fuel to power battery electric vehicles (BEVs). These vehicles store the electricity in a battery, which then powers the vehicle's wheels through an electric motor. The batteries have a limited storage capacity, and once the charge has been used up they must be recharged by being plugged into an electrical source. Examples of these vehicles are the battery-powered escort courtesy cars common in airport terminals.

Ethanol

Ethanol is a renewable fuel made from biomass (various plant materials). Ethanol contains the same chemical compound (C_2H_5OH) found in alcoholic beverages, which is why it is also known as ethyl alcohol or grain alcohol. Presently, nearly half of the gasoline produced in the United States contains ethanol in a low-level blend in order to oxygenate the fuel and reduce air pollution. According to the DOE, studies have estimated that ethanol and other biofuels could replace 30 percent or more of the U.S. gasoline demand by 2030.

It takes several steps to produce ethanol. Initially, biomass feedstock must be grown. Ethanol is a clear, colorless liquid and can be produced from either starch- or sugar-based feedstock such as corn grain or sugarcane. In the United States, ethanol is primarily produced from corn grain. In countries such as Brazil it is produced from sugarcane. It can also be produced from cellulosic feedstock such as grass, wood, crop residues, or old newspapers, but the process is more involved and complicated.

Plants contain the cellulosic materials cellulose and hemicellulose. These complex polymers are what form the structure of plant stalks, leaves, trunks, branches, and husks. They are also found in products made from plants, such as sugar. In order to make ethanol from cellulosic feedstock the materials must be broken down into their component sugars for fermentation to ethanol in a process called biochemical conversion. Cellulosic feedstock can also be converted into ethanol in a process called biochemical conversion. Cellulosic ethanol conversion processes are a major focus of current DOE research.

According to scientists at the DOE, ethanol works very well in internal combustion engines. Historically, Henry Ford and other early automakers believed ethanol would be the world's primary fuel source (before gasoline became so readily available).

As a comparison, a pure gallon of ethanol contains 34 percent less energy than a gallon of gasoline. It is also a high-octane fuel. Low-octane gasoline can be blended with 10 percent ethanol to achieve the standard 87 octane requirement. Ethanol is the principal component today of 85 percent ethanol and 15 percent gasoline (E85).

The DOE has identified several benefits of ethanol. It is a renewable, largely domestic transportation fuel. When it is used as a low-level blend such as E10 (10 percent ethanol, 90 percent gasoline) or high-level such as E85 (85 percent ethanol, 15 percent gasoline), ethanol still helps reduce the U.S. dependence on foreign oil and combats global warming. In the United States, ethanol is produced almost entirely from domestic crops.

Another major positive is that the CO_2 released when ethanol is burned is balanced by the CO_2 captured when the crops are grown to make ethanol. Based on data from Argonne National Laboratory, on a life-cycle basis, corn-based ethanol production and use reduces greenhouse gas emissions by up to 52 percent compared to gasoline production and use. They project that cellulosic ethanol use could reduce greenhouse gases (GHGs) by as much as 86 percent. Argonne National Laboratory is the DOE's largest research center, located in Chicago.

Ethanol is also completely biodegradable and, if spilled, poses much less of an environmental threat than does petroleum to surface and groundwater. As an example of this, after the sinking of the *Bow*

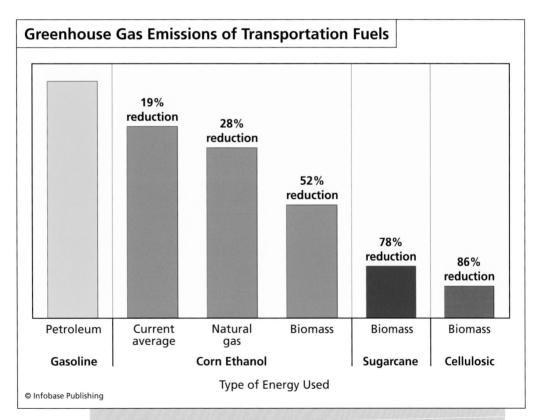

Different transportation fuels have different greenhouse gas reduction capabilities.

Mariner off the Virginia coast in February 2004, U.S. officials noted the 3.2 million gallon (12 million liter) cargo of industrial ethanol had dissipated quickly and did not pose an environmental threat to humans or marine life.

For more than 20 years, the DOE has been heavily involved in research and development to identify and develop promising energy crops. Their prominent research areas include an integrated analysis of biomass resources, feedstock sustainability, and feedstock systems engineering. The U.S. Department of Agriculture (USDA) is also heavily involved in ethanol feedstock research and development. Two specific fields they are currently focused on include selective breeding and genetic engineering to improve feedstock crop yields, developing sustainable approaches to feedstock production, and investigating the environmental and economic impacts of feedstock crops on farmland.

The Biomass Research and Development Initiative is a DOE/USDA combined effort to coordinate and accelerate all federal biobased products and bioenergy research and development. The initiative funds ethanol feedstock production research and development. The DOE office of science also supports fundamental research on ethanol feedstock, including the development of optimized energy crops. It has formed three Bioenergy Research Centers to accelerate basic research on cellulostic ethanol and other biofuels.

Methanol

Methanol, also known as wood alcohol, is considered an alternative fuel under the Energy Policy Act of 1992. Currently, most of the methanol production is accomplished through a process that uses natural gas as a feedstock. Methanol can be used to make methyl tertiary-butyl ether (MTBE), an oxygenate that is blended with gasoline to enhance octane and create cleaner burning fuel. MTBE production and use has declined in recent years, however, because it was responsible for some groundwater contamination.

As an engine fuel, methanol has similar chemical and physical characteristics as ethanol. Methanol is methane with one hydrogen molecule replaced by a hydroxyl radical. In the formation process, steam reforms

Ethanol and methanol fuel pumps for refueling alternative fueled vehicles *(Warren Gretz, DOE/NREL)*

natural gas to create a synthesis gas, which is then fed into a reactor vessel in the presence of a catalyst. This process produces methanol and water vapor. Even though a variety of feedstock can be used to create methanol, with today's economy, the use of natural gas is the preferred production method.

There are several advantages to methanol's physical and chemical characteristics for use as an alternative fuel:

- It has a relatively low production cost.
- It has a lower risk of flammability compared to traditional gasoline.
- It can be manufactured from a variety of carbon-based feedstock.
- It can help reduce the U.S. dependence on foreign oil.
- It can be converted into hydrogen.

Researchers are currently trying to find a way to use methanol in fuel cell vehicles for use in the future.

Hydrogen

Hydrogen is the latest energy source that science is looking toward with the expectation that it could completely revolutionize not only the alternative fuel industry, but also possibly the entire energy system. Hydrogen is an energy carrier, not an energy source. Energy is required to separate it from other compounds. Once produced, hydrogen stores energy until it is delivered in a usable form, such as hydrogen gas delivered into a fuel cell. Hydrogen is the lightest element on the periodic table and the most abundant element in the universe. It makes up roughly 90 percent of the universe by weight. Hydrogen is the lightest of the gases, and when it is burned, its only waste product is water. Because of its abundance, simplicity, efficiency, and lack of toxic emissions, it is viewed as the perfect fuel in the face of global warming.

Hydrogen can be produced from fossil fuels, nuclear energy, biomass, and by electrolyzing water. The environmental impact and energy efficiency of hydrogen depends on how it is produced. If hydrogen is produced with renewable energy and used in fuel cell vehicles, it is the one alternative fuel that holds the promise of eventually obtaining a virtually pollution-free transportation system network and a long-awaited freedom from the U.S. crippling dependence on foreign oil in an extremely politically unbalanced world.

At the Earth's surface temperature and pressure, hydrogen is colorless. It is rarely found alone in the natural world, however. It is usually bonded with other elements. Only small amounts are actually present in the Earth's atmosphere. Hydrogen exists in enormous quantities in water, in hydrocarbons, and in other organic matter. Efficiently producing hydrogen from existing compounds is the major hurdle that researchers are facing today.

According to the DOE, natural gas reforming using steam accounts for about 95 percent of the approximately 9 million tons (8 million metric tons) of hydrogen produced in the United States each year. This level of hydrogen production could fuel more than 34 million cars. The major hydrogen-producing states today are California, Louisiana, and Texas. Almost all of the hydrogen produced in the United States is used for refining petroleum, treating metals, producing fertilizer, and processing foods. It has also been used extensively by NASA for space flight since the 1950s.

Hydrogen is also used to fuel internal combustion engines and fuel cells, which can then power low- or zero-emission vehicles such as fuel cell vehicles. Today, there are major research and development efforts going on to make this a viable and conveniently accessible technology for the public. For years, hydrogen technology has been looked at as the ideal technology to supply the world with energy and still win the battle against global warming.

Hydrogen is considered an alternative fuel under the Energy Policy Act of 1992. Its chief interest as an alternative transportation fuel is attributed to three factors: (1) It is clean burning; (2) it can be produced domestically, lowering/eliminating dependence on foreign countries; and (3) it has an extremely high efficiency potential (up to three times more efficient than today's gasoline-powered vehicles).

There is still research that needs to be done, however. The energy in 2.2 pounds (1 kg) of hydrogen is equivalent to the energy in one gallon (3.8 liters) of standard automobile gasoline. In order for a vehicle to travel 300 miles (483 km) a fuel cell would have to store 11 to 29 pounds (5 to 13 kg) of hydrogen. Because hydrogen has a low volumetric energy density (a small amount of energy by volume compared with fuels such as gasoline), it would require a tank larger than a car's trunk to store it all. Therefore, it is not quite practical yet; advanced technologies are needed to reduce storage space and weight.

Hydrogen-storage technologies currently being researched encompass high-pressure tanks with gaseous hydrogen compressed up to 10,000 pounds per square inch, cryogenic liquid hydrogen cooled –423°F (–253°C) in insulated tanks, and chemical bonding of hydrogen with another material (such as metal hydrides).

Natural Gas

Natural gas is a mixture of hydrocarbons, mostly methane. When it is delivered through the pipeline system in the United States, it also contains hydrocarbons such as ethane and propane and other gases such as helium, CO_2, nitrogen, hydrogen sulfide, and water vapor.

Natural gas has a high octane rating and works well in spark-ignited internal combustion engines. It is one of the cleaner fossil fuels, is non-toxic, noncorrosive, and noncarcinogenic. It also has the distinction of not polluting soil, groundwater, or surface water. The bulk of usable natural gas is obtained from oil and gas wells. Smaller amounts can be obtained from other sources such as synthetic gas, landfill gas, other biogas resources, and gas by-products from coal.

Roughly one-fourth of the energy used in the United States comes from natural gas. Of that amount one-third goes to commercial and residential uses, one-third to industrial uses, and one-third to electrical power production. Only about one-tenth of 1 percent is currently used for transportation fuel. Due to the fuel's gaseous nature, it has to be stored on board a vehicle in either a compressed gaseous (compressed natural gas, CNG) or liquefied (liquefied natural gas, LNG) state.

CNG must be stored on board a vehicle in tanks at high pressure—up to 3,600 pounds per square inch (psi). A CNG-powered vehicle gets about the same fuel economy as a conventional gasoline vehicle on a gasoline gallon equivalent (GGE) basis. A GGE is the amount of alternative fuel that contains the same amount of energy as a gallon of gasoline. A GGE equals about 5.7 pounds (2.6 kg) of CNG.

In order to store more energy on board a vehicle in an even smaller volume, natural gas can be liquefied. To produce LNG, natural gas is purified and condensed into liquid by cooling it to -260°F (-162°C). At atmospheric pressure, LNG occupies only 1/600 the volume of natural gas in its vapor form. A GGE equals about 1.5 gallons (5.7 liters) of LNG. Because it must be kept at such cold temperatures, LNG is stored in double-wall, vacuum-insulated pressure vessels. LNG fuel systems are mainly used with heavy-duty vehicles.

Most natural gas is found as a fossil fuel formed over millions of years by the action of heat and pressure on organic material like ancient plants and animals. It can also be found—in much smaller quantities—in landfill gas and sewage treatment areas.

Natural gas is released from where it is buried beneath the Earth's surface. It is usually located in subsurface porous rock reservoirs. Gas streams produced from oil and gas reservoirs can contain natural gas, liquids, and other materials. Processing is required to separate the gas from free liquids such as crude oil, hydrocarbon condensate, and water. The separated gas is further processed after that. The DOE has identified some benefits of the use of natural gas as an alternative fuel: It is domestically available and inherently clean burning. It gives the United States more energy security because it is available locally.

Natural gas vehicles and infrastructure development (the filling stations, etc.) can also facilitate the transition to hydrogen technology and fuel cell vehicles. According to the DOE, with the highest hydrogen-to-carbon ratio of any energy source, natural gas is an efficient source of hydrogen—it is the number one source of commercial hydrogen used in the United States. The vast network of natural gas transmission lines offers the potential for convenient transportation of natural gas to future refueling stations that reform hydrogen from the gas. The DOE has identified important similarities between natural gas and hydrogen technologies that make the lessons learned from natural gas technology an aid to the future transition from conventional liquid fuels to gaseous hydrogen fuel. Their similarities include fuel storage, fueling methods, station site locations, facilities, and public acceptability.

Propane

Propane—also known as liquefied petroleum gas (LPG), or autogas in Europe, is a three-carbon alkane gas. Stored under pressure inside a tank, propane turns into a colorless, odorless liquid. As pressure is released, the liquid propane vaporizes and turns into gas that is used for combustion. Propane has a high octane rating. It is nontoxic and presents no threat to soil, surface water, or groundwater.

Propane is produced as a by-product of natural gas processing and crude oil refining. It accounts for only about 2 percent of the energy used in the United States and is used to heat homes (usually in more remote, or rural, areas where gas lines have not been run), cooking, refrigerating food, clothes drying, powering farm and industrial equipment, drying corn, and heating barbecues. The chemical industry uses propane as a raw material for making plastics and other compounds. Less than 2 percent of propane consumption is used for transportation fuel.

The idea of using liquefied petroleum gas as an alternative transportation fuel comes mainly from the fact that it has domestic availability.

Because of its high energy density and its clean burning qualities, it is also the most commonly used alternative transportation fuel and the third most used vehicle fuel (behind gasoline and diesel). When propane is used as a vehicle fuel, it can be run as a mixture of propane with smaller amounts of other gases. According to the Gas Processors Association's HD-5 specification for propane as a transportation fuel, it must consist of 90 percent propane, no more than 5 percent propylene, and 5 percent other gases, mainly butane and butylene.

Propane—a gas at normal pressure and temperature—is stored on board vehicles in a tank pressurized at 300 pounds psi—equal to a pressure about twice that of an inflated truck tire. A gallon of propane has about 25 percent less energy than a gallon of gasoline.

NEW FUEL TECHNOLOGY

Researchers are busy today trying to identify new emerging vehicle alternative fuels. Several fuels are just in the beginning stages of development. According to the DOE, each one of them promises benefits in the forms of increased energy, increased national security, reduced emissions, higher performance, and economic stimulation. Most of the fuels discussed in this section are considered alternative fuels under the Energy Policy Act of 1992. They include biobutanol, biogas, biomass to liquids, coal to liquids, gas to liquids, hydrogenation-derived renewable diesel, P-series fuels, ultra-low sulfur diesel, and green charcoal.

Biobutanol

Biobutanol is a 4-carbon alcohol (butyl alcohol), produced from biomass feedstock. Its primary use today is as an industrial solvent in products like enamels and lacquers. Similar to ethanol, biobutanol is a liquid alcohol fuel that can be used in today's gasoline-powered internal combustion engines. Its chemical properties make it a good fuel to blend with gasoline. It is also compatible with ethanol blending and can improve the blending of ethanol with gasoline. The energy content of biobutanol is 10 to 20 percent lower than that of gasoline.

According to the EPA, biobutanol can be blended as an oxygenate with gasoline in concentrations up to 11.5 percent by volume. Blends of 85 percent or more biobutanol with gasoline are required to qualify as an alternative fuel. Today, a company called Butyl Fuel, LLC, through a DOE Small Business Technology Transfer grant, is working on developing a process aimed at making biobutanol production economically competitive with petrochemical production processes. Butyl Fuel's current plans are to market its biobutanol as a solvent first and then market it as an alternative fuel in the future. To date, there is no infrastructure in place for fueling vehicles with biobutanol, but because biobutanol does not corrode pipes or contaminate water as ethanol can, researchers expect that biobutanol will be able to be distributed through existing gasoline infrastructure, including existing pipelines.

There are several benefits of biobutanol. The DOE has identified the following:

- It can be produced domestically from several homegrown types of feedstock.
- Its domestic production can create more jobs in the United States.
- Greenhouse gas emissions are reduced because the CO_2 captured when the feedstock crops are grown balances the CO_2 released when the biobutanol is burned.
- It is easily blended with gasoline for use in vehicles.
- Its energy density is only 10 to 20 percent lower than gasoline's.
- It is compatible with the current gasoline distribution infrastructure and would not require new or modified pipelines, blending facilities, storage tanks, or retail station pumps.
- It is compatible with ethanol blending and can improve the blending of ethanol with gasoline.
- It can be produced using existing ethanol production facilities with relatively minor modifications.

Biobutanol is also currently being researched by other industries and government groups. The USDA Agricultural Research Service (ARS) is studying biobutanol production as a part of a study in its bioprocess technologies for production of biofuels from lignocellulosic biomass.

Biogas

Biogas is a gaseous product of the anaerobic digestion of organic matter, from sources such as sewage sludge, agricultural wastes, industrial wastes, animal by-products, and municipal solid wastes. In landfills, anaerobic digestion of wastes occurs naturally. Gas collection is viable in landfills that are at least 40 feet (12 m) deep and contain at least 1 million tons (907,185 metric tons) of waste. It consists of 50 to 80 percent methane, 20 to 50 percent CO_2, and traces of hydrogen, nitrogen, and carbon monoxide. It is sometimes referred to as swamp gas, landfill gas, or digester gas. When its composition is upgraded to a higher standard of purity, it is referred to as renewable natural gas.

After biogas is produced and extracted, it has to be upgraded for pipeline distribution or use as a vehicle fuel. This process requires that the methane proportion be increased and the CO_2 contaminants be decreased. The International Energy Agency (IEA) estimated that in 2005 185 anaerobic digestion plants had the capacity to process 5.5 million tons (5 million metric tons) of municipal solid and organic industrial waste to generate 600 megawatts (MW) of electricity. In a report issued by the CIVITAS Initiative, it was estimated that European biogas production could satisfy up to 20 percent of Europe's natural gas consumption. A report titled "Natural Gas Vehicles for America" cites a 1998 study estimating that the biogas potential at that time from landfills, animal waste, and sewage was equivalent to 6 percent of U.S. natural gas consumption or 10 billion gasoline gallon equivalents of transportation fuel. This calculates into about 7 percent of the 2006 U.S. gasoline consumption.

Biogas is used for many different purposes. In rural communities, it is used for household cooking and lighting. Large-scale digesters provide biogas for heat and steam, electricity production, chemical production, and vehicle fuel. Once biogas is upgraded to the required level of purity and compressed or liquefied, biogas can be used as an alternative vehicle fuel in the same form as conventionally derived natural gas (CNG or LNG).

A 2007 DOE report estimated that 12,000 vehicles worldwide are using biogas. They predict that by 2010 there may be 70,000 biogas-fueled vehicles on the roads worldwide. The majority of these vehicles exist in Europe, with Sweden claiming that more than half of the gas used in its 11,500 natural gas vehicles is biogas. Biogas vehicle activity has not been as high in the United States. DOE's research and development–sponsored projects have been working on further development of biogas technologies. The DOE has identified several benefits of biogas as an alternative fuel. Its use represents increased energy security for the nation as well as a conduit to pave the way for fuel cell vehicles in the future. It also improves public health and the environment through reduced vehicle emissions and off-sets the use of nonrenewable resources such as coal, oil, and natural gas. It reduces greenhouse gas emissions, treats waste disposal naturally, requires less land area, and reduces the amount of material that must be landfilled. Its production also creates jobs and benefits for local economies.

Biomass to Liquids (BTL)

Biomass to liquids describes processes for converting biomass into a range of liquid fuels such as gasoline, diesel, and petroleum refinery feedstock. These processes are different from the enzymatic/fermentation processes and processes that use only part of a biomass feedstock, such as the processes that produce ethanol, biobutanol, and biodiesel. Currently, the major biomass to liquids production processes are gas-to-liquids, which involve the conversion of biomass into gas and then into liquids, and pyrolysis, involving the decomposition of biomass in the absence of oxygen to produce a liquid oil.

Biomass to liquids processes have the potential to produce a wide range of fuels and chemicals. These fuels include gasoline, diesel, and ethanol. A major benefit of these fuels is their compatibility with currently existing vehicle technologies and fuel distribution systems. Biomass-derived gasoline and diesel could be transported through existing pipelines, dispensed at existing fueling stations, and used to fuel today's gasoline- and diesel-powered vehicles. Another major benefit is that these fuels reduce regulated exhaust emissions from a variety of diesel engines and vehicles, and their near-zero sulfur content enables the use of advanced emission control devices.

Coal to Liquids (CTL)

Coal to liquids is the process of converting coal into liquid fuels such as diesel and gasoline. The principal method available today is called the

Fischer-Tropsch process, which is a two-step process: It first converts the coal into gas, and then into liquid. There are several other processes that are able to directly convert coal into liquids—a process called liquefaction—but they are not as common.

Coal to liquids has the ability to produce a number of useful fuels and chemicals, including transportation fuels. In addition to diesel and gasoline, they can also produce methanol. The largest benefit of this technology is that the resulting fuel is compatible with currently existing vehicle technology and fuel distribution systems. In addition, coal-derived gasoline and diesel could be transported through the existing pipeline infrastructure, dispensed at existing fueling stations, and used in vehicles without any modifications.

The DOE has identified the following benefits to pursuing research and development with the coal to liquid technology: The fuel can be used directly in existing vehicles, it is compatible with existing infrastructure (pipelines, storage tanks, and retail station pumps); and the fuel provides similar or better vehicle performance than conventional diesel. In addition, the Fischer-Tropsch diesel reduces regulated exhaust emissions from a variety of diesel engines and vehicles. The near-zero sulfur content of the fuels allows the use of advance emission control devices.

The EPA has run their own tests on the Fischer-Tropsch diesel and determined the coal to liquid diesel has benefits over traditional diesel, including the fact that it was cleaner burning, with less nitrogen oxide, little to no particulate emissions, and lower in hydrocarbon and carbon monoxide emissions.

Gas to Liquids

Gas to liquids is the process of converting natural gas into liquid fuels such as diesel, methanol, and gasoline. The principal production method is the Fischer-Tropsch process. This conversion process can produce a range of fuels and chemicals. Similar to the coal to liquid, a major benefit of the gas to liquid is its compatibility with existing vehicle technologies and fuel distribution infrastructure.

To date, tests that have been run on the diesel created using the Fischer-Tropsch method provide similar or even better vehicle performance than conventional diesel does. These fuels can also be produced using natural gas reserves that are uneconomical to recover using other methods, such as stranded reserves. According to a report in the *Petroleum Economist* in May 2002, there are about 800 small, undeveloped fields (called stranded reserves) that are potential candidates for Fischer-Tropsch gas to liquids projects of up to around 10,000 barrels a day. If stranded reserves are used to produce liquid fuels, it reduces the need to flare natural gas, which results in reduced greenhouse gas emissions.

Hydrogenation-Derived Renewable Diesel

Hydrogenation-derived renewable diesel (HDRD) is the product of fats or vegetable oils—either alone or blended with petroleum—that have been refined in an oil refinery. HDRD that is produced this way is sometimes referred to as second-generation biodiesel. Not much work has been done on it, but researchers expect HDRD will be able to either substitute directly for or blend in any proportion with petroleum-based diesel, without modification to vehicle engines or the fueling infrastructure. HDRD's ultralow sulfur content and high cetane number (a measure of the combustion quality of diesel fuel) will provide both vehicle performance and emissions benefits. The technology is not widely available today, but it is gaining recognition.

Researchers see many benefits resulting from its development and future implementation. For example, it should be able to be produced domestically creating U.S. jobs; it reduces greenhouse gas emissions because the CO_2 is captured when the feedstock crops are grown, balancing the CO_2 release when the fuel is burned; it should be able to be used in today's diesel-powered vehicles; it should be fully compatible with current infrastructure (pipelines, fueling stations and storage); and it can be produced using existing oil refinery capacity and does not require extensive new facilities. It should also provide, similar, or even better, performance over commercial diesel. It also has an ultralow sulfur content and should be able to work with advanced emission control devices on all vehicles.

P-Series

P-Series fuel is a blend of natural gas liquids, ethanol, and the biomass-derived cosolvent methyltetrahydrofuran (MeTHF). P-Series fuels are a clear, colorless, 89–93 octane, liquid blends that are formulated to be used in *flexible fuel vehicles.* P-Series fuel can be used alone or mixed with gasoline in any proportion inside a flex fuel vehicle's fuel tank. Currently, the P-Series is not yet being produced in large quantities and is not widely used.

Ultralow Sulfur Diesel

Ultralow sulfur diesel is diesel fuel with 15 parts per million (ppm) or lower sulfur content. The low sulfur content enables the use of advanced emission control technologies on both light-duty and heavy-duty diesel vehicles. Most of the highway diesel fuel refined in or imported into the United States is required to be ultralow sulfur diesel nationwide by 2010. Today, most of the ultralow sulfur diesel is produced from petroleum. In the future, it will be possible to make it from the following processes: biomass to liquids, coal to liquids, and gas to liquids. The EPA has identified the following major benefits:

- Ultralow sulfur diesel can use catalytic converters and particulate traps that nearly eliminate emissions of nitrogen oxides and particulate matter, all pollutants related to serious health problems.
- Emission reductions from the use of clean diesel will be equivalent to removing the pollution from more than 90 percent of today's trucks and buses, when the current heavy-duty vehicle fleet has been completely replaced in 2030.
- Diesel engines are 20 to 40 percent more efficient than comparable gasoline engines.
- Ultralow sulfur diesel uses existing fueling infrastructure and works with existing engine and vehicle technologies.
- Replacing some gasoline vehicles with diesel vehicles will result in reduced U.S. petroleum fuel use and greenhouse gas emissions.

The Energy Policy Act of 1992 was passed by Congress to reduce the nation's dependence on imported petroleum by requiring certain fleets to acquire alternative fuel vehicles, which are capable of operating on nonpetroleum fuels. Ultralow sulfur diesels will assist in working toward that goal.

Green Charcoal

The development of cooking fuels using biomass has created another type of fuel—that of green charcoal. This process involves a continuous process of pyrolysis of vegetable waste (agricultural residues, renewable, wild-growth biomass) and transforms it into a product referred to as green charcoal. This newly created domestic fuel performs the same as charcoal made from wood, but at half the cost. Used in Africa, this fuel has many more uses than cooking; it also represents a freedom from being held hostage to a scarcity of resources and the long distances to and great costs of available fuels in Africa.

Today, more than 2 million people worldwide face domestic energy shortages. In many parts of Africa, Latin America, and Asia, wood is becoming more scarce, and modern energy supplies are difficult to come by or nonexistent. In the Sahel region, inhabitants have to walk about 12 miles (20 km) a day just to find a household supply of wood. In the small villages, families are forced to spend up to one-third of their income on wood or charcoal. As the villagers gather wood, they steadily deforest the area increasing the ill effects of drought and deforestation, which then leads to climate change.

In an effort to halt deforestation in the savanna zones, a renewable replacement substance to effectively use as an energy source was introduced to the people in the region. A process was developed to carbonize vegetative material into pellets or briquettes. Savanna weeds, reeds, and various types of straw (wheat, rice, or maize), cotton stems, rice husk, coffee husk, bamboo, or any plant with a sufficient lignin content can be used to produce green charcoal. The advantage to using green charcoal is that it preserves forests, keeping them from being deforested. Green charcoal also eliminates methane emissions. The process of producing the briquettes does not release any greenhouse gases.

Another reason for using green charcoal is to avoid the buildup of soot. According to an April 16, 2009, article in the *New York Times* by Elisabeth Rosenthal, soot (also called black carbon) builds up in tens of thousands of villages in the developing world, based on research con-

Green charcoal briquettes are made from biomass, sawdust, and shredded and soaked paper in Bukava and the Virunga Mountains in Congo. This alternative fuel is used instead of traditional charcoal to protect the mountain gorillas of the region (obtaining charcoal in the region endangers the gorilla). The biomass briquettes burn charcoal. They are mixed and patted into molds until they dry into briquettes (top image). Once they are ready to burn, their heat output is similar to charcoal, enabling the villagers to cook their food efficiently and sustainably. *(Wildlife Direct Organization)*

ducted by Dr. Veerabhadran Ramanathan, professor of climate science at Scripps Institute of Oceanography and one of the world's leading climate scientists. It is actually a pollutant and a major (although previously ignored) source of global climate change.

Ramanathan says that black carbon has recently emerged as a serious contributor—responsible for 18 percent of the global warming that has occurred, compared to a 40 percent contribution from CO_2. Focusing on black carbon emissions and reducing them is currently being viewed by scientists such as Ramanathan as an inexpensive, easy, relatively fast way to slow global warming, especially in the near term. One way to accomplish this goal is to replace the common primitive cooking stoves found in most homes in developing countries with a modern version that emits much less soot.

According to Ramanathan, "It is clear to any person who cares about climate change that these [replacement stoves] will have a huge impact on the global environment. In terms of climate change, we're driving fast toward a cliff, and this could buy us time."

Another advantage to this is that because soot has a fairly short life span (a few weeks), decreasing the input of soot in the atmosphere would have an immediate effect in combating global warming.

The discovery of black carbon's influence is so new that it was not even mentioned in the IPCC's 2007 report. It has been through recent research by institutions such as Scripps and NASA that black carbon has recently become better understood. It is now believed that black carbon could account for as much as half of the current Arctic warming. While soot does not travel globally like CO_2, it does travel. Soot from India has been found in the Maldives and on the Tibetan Plateau. From the United States, it travels to the Arctic. Professor Syed Iqbal Hasnain, a glacier specialist in India, has predicted that the Himalayan glaciers will most likely lose 75 percent of their mass by 2020 because of black carbon deposition.

In an effort to take action on this, the U.S. Congress introduced a bill in March 2009 that would require the EPA to specifically regulate black carbon and direct aid to black carbon reduction projects abroad. This major effort would include providing modern cookstoves (efficient, low soot-producing) to 20 million homes in developing countries where black carbon pollution poses the biggest problems. These new stoves cost about $20 and use solar power, making them much more efficient while reducing soot levels more than 90 percent. The solar stoves do not use wood or biomass. Other new stove options burn cleaner. In March 2009, a cookstove project called Surya was initiated. It began market-testing six different styles of cookstoves in small villages in India in an effort to begin the conversion of stoves to reduce black carbon pollution.

Research scientists are busy developing alternative and advanced fuels and are also busy looking for options beyond fossil fuels to supply energy for the future. If significant advances are not made within the next decade, there will be little hope of offsetting the permanent destruction from global warming. Research scientists—and policy makers—are up against the clock, which is ticking fast.

Green Technology: Tomorrow's Cars Today

One of the most effective ways people can be empowered to fight global warming is through their choice of vehicle. The types of cars available on the market today offer a wide range of options—from inefficient gas-guzzlers to fuel-efficient models. Each person's choice makes a difference in the picture of the progression of global warming. This chapter looks at the latest green vehicle technology and the direction research is going to help solve global warming.

DRIVING INTO THE FUTURE

It is often said that Americans have been in love with their cars ever since Henry Ford turned cars into a must-have item. Interestingly, when Ford's Model T was introduced in 1908, it got 28.5 miles per gallon (MPG) fuel efficiency. Since then, even though technology has improved and cars today have a wide selection of features and go farther and faster, when it comes to fuel efficiency, technology has digressed.

Vehicles have not evolved with fuel efficiency as a goal. In response to the oil shortage crisis in the 1970s, the United States was determined to begin producing cars that doubled the mileage of existing cars. Since then, however, with the pressure off, the worry about fuel efficiency went by the wayside and the focus shifted instead to performance cars such as SUVs. By 2005, most of the cars on the highways in the United States were less fuel efficient than those on the road in the 1980s.

According to the Environmental Defense Fund (EDF), if the exhaust coming from a car had an actual weight, an average household with two medium-sized sedans would emit more than 20,000 pounds (10 tons) of carbon dioxide (CO_2) a year. Even worse, SUVs emit up to 40 percent more than smaller cars.

Today, fuel efficiency has once again become a concern for two reasons: global warming and the dependence on foreign sources. A gallon of gasoline weighs slightly more than 6 pounds (2.7 kg). When it is burned as fuel in a vehicle, the carbon in it combines with oxygen and produces approximately 19 pounds (8.6 kg) of CO_2. If the energy is added in that was expended in making and distributing the fuel, the total global warming pollution is about 25 pounds (11 kg) of CO_2 per gallon. To illustrate the impact, a car that gets 21 MPG (34 km/g) and is driven 30 miles (48 km) a day uses 1.4 gallons each day and emits 35.7 pounds (16 kg) of CO_2 every day. When multiplied by the millions of cars that are driven in the United States each day, this adds up, and the United States is just one country in the world. This means that 1 million cars emit the equivalent of 35.7 million pounds of CO_2 every day, 2 million cars contribute 71.4 million pounds of CO_2 every day, and 3 million cars contribute 107.1 million pounds of CO_2 every day.

The table on the opposite page illustrates what the true costs of lower fuel efficiency add up to and why it is important to take action now toward a greener tomorrow.

For the average American who owns a car, driving is one of the top two daily pollution-causing activities (electricity use is the other one). Because of this, by choosing a greener vehicle, it is one way that a person can make a significant difference in the fight against global warming. According to the EDF, vehicle choice is one of the most powerful

The Annual Cost of Lower Fuel Efficiency			
AVERAGE GAS MILEAGE	AVERAGE FUEL USED (BASED ON 12,000 MILES PER YEAR)	APPROXIMATE GREENHOUSE GAS POLLUTION	APPROXIMATE COST (BASED ON $2.30/GALLON)
50 MPG	240 gallons	2.7 tons/year	$552
40 MPG	300 gallons	3.4 tons/year	$690
30 MPG	400 gallons	4.5 tons/year	$920
25 MPG	480 gallons	5.4 tons/year	$1,104
20 MPG	600 gallons	6.8 tons/year	$1,380
15 MPG	800 gallons	9 tons/year	$1,840
10 MPG	1,200 gallons	13.6 tons/year	$2,760

Source: Argonne National Laboratory

choices a person can make, and there is a triple benefit associated with it: (1) It protects the climate; (2) it reduces the U.S. dependence on oil; and (3) it saves money at the pump.

Population growth and the spread of suburban areas have put even more cars on the road and mean greater distances driven each day. These trends combined with inefficient mileage spell environmental disaster. One way to combat this dangerous trend is with green technology in the auto industry—the advent of the hybrid cars, electric vehicle options, flexible fuels, fuel cells, plug-in cars, and other cutting-edge forms of transportation technology currently being developed.

HYBRIDS

Hybrids combine a small combustion engine with an electric motor and battery. The two technologies can be combined to reduce fuel consumption and tailpipe emissions. Most of the hybrids on the road today complement their gas engines by charging a battery while braking—a concept called regenerative braking.

Alternative fuel vessels are doing their part to cut traffic congestion and emissions in Fort Lauderdale, Florida. The hybrid-electric
waterbus runs on electricity and B20 bio-diesel, consuming just half
the fuel per day as a similar diesel-powered boat. Approximately
400,000 passengers per year park their cars and commute around
Fort Lauderdale via water taxi. *(Water Taxi, DOE/NREL)*

Engines that run on diesel or other alternative fuels can also be used
in hybrids. A hybrid drive is fully scalable, which means that the drive
can be used to power everything from small commuter cars to large
buses and watercraft. The technology can even work on locomotives.
Hybrids get more miles per gallon than most non-hybrids; they also
usually have very low tailpipe emissions.

Hybrids are able to reduce smog pollution by 90 percent compared
with the cleanest conventional vehicles on the road today. For example,
the Toyota Prius, introduced in 1997, achieves a 90 percent reduction
in smog-forming pollutants over the current national average. Because
hybrids do have an internal combustion engine, however, they will
never be able to achieve zero emissions. They do consume much less
fuel. They do cut global warming emissions by a third to a half. Models
currently under development should be able to cut even more.

By combining gasoline and electric power, hybrids have the same or greater range than traditional combustion engines. For example, the 2009 Honda Insight goes about 700 miles (1,127 km) on a single tank of gas. The 2009 Toyota Prius gets 48 miles (77 km) per gallon in highway driving and 45 miles (72 km) per gallon in city driving. For driving, hybrids have the same or better performance than their traditional counterparts. As public support catches on, more hybrids will be introduced. The initial cost of hybrids is slightly higher, but with fuel savings over the length of the car's lifetime, the costs are competitively priced. Another major benefit of hybrids is that purchasers may qualify for a federal income tax credit.

ELECTRIC VEHICLES

A battery-electric vehicle (BEV) uses electricity stored in its battery pack to power an electric motor that turns its wheels. The battery pack is recharged by connecting it (plugging it in) to a wall socket or other electrical source, such as a solar panel.

Because these vehicles use electricity as the fuel source, there are no emissions from its tailpipe when recharging the electricity, and it costs pennies (compared to dollars at the pump for gasoline-powered vehicles). Like all technology, there are both benefits and limitations to battery-electric vehicles. Since BEVs do not have tailpipes, they do not produce any tailpipe emissions. They do recharge, however, using electricity generated at power plants that emit global warming and smog-forming pollutants. If an electric vehicle is charged at a facility that strictly uses renewable energy such as solar power, hydropower, or wind power, then electric car technology is completely green. If the vehicle is recharged at a facility whose electricity is generated from power plants that use fossil fuels, they are still up to 99 percent cleaner than conventional vehicles and can cut global warming emissions by as much as 70 percent. They are energy efficient. The electric motors convert 75 percent of the chemical energy from the batteries to power the wheels. As a comparison, internal combustion engines convert only 20 percent of the energy stored in gasoline.

BEVs are more expensive to purchase than standard cars. This is largely attributed to the fact that their advanced battery packs are

Electric car *(Tom Brewster, DOE/NREL)*

expensive to produce. The positive side is that they only cost about one-third the price of refueling as a gasoline-powered vehicle. BEVs also offer quiet, smooth driving experiences. They can travel 50 to 100 miles (80 to 161 km) per charge depending on the battery type and driving conditions. Electric vehicles reduce energy dependency because the energy supply is solely domestic. It takes four to eight hours to fully recharge the battery.

FLEXIBLE FUEL VEHICLES

Flexible fuel vehicles (FFVs) are designed to run on gasoline or a blend of up to 85 percent ethanol. They are similar to gasoline models, having only a few engine and fuel system modifications. FFVs have been produced since the 1980s. The opinion of the Union of Concerned Scientists (UCS), when discussing global warming and the environment, is that while there may be potential benefits from getting more FFVs out on the road, the benefits are not worth any increase in oil dependency.

What the UCS supports instead is the capability of driving cars that run on alternative fuels as long as the main fuel choices are not fossil fuels—they need to be either biomass fuels, domestic fuels, and advanced, alternative fuels that are good for the environment, national security, and the economy.

One of the UCS's main concerns about the FFV program is what they refer to as the dual-fuel loophole. The dual-fuel loophole allows manufacturers to earn credits toward meeting federal fuel economy standards by producing vehicles that are able to run on both petroleum and an alternative fuel, even if they never actually use the alternative fuel. They claim that the way the program is currently set up, auto manufacturers can sell fleets of vehicles that fall short of federal fuel economy targets. By their calculations, even back in 2004, the loophole was already increasing U.S. oil dependence by 80,000 barrels per day. The best approach to this dilemma according to the UCS is to make sure that auto manufacturers are ensuring that the fuel tank designed to operate on the 85 percent ethanol fuel is reliable so that it is what the vehicle owner uses.

FUEL CELLS

There are currently several types of fuel cells. The type most often used in vehicles is polymer electrolyte membrane (PEM) fuel cells—also called proton exchange membrane fuel cells. A PEM fuel cell uses hydrogen fuel and oxygen from the air to produce electricity.

Hydrogen fuel is channeled through field flow plates to the anode on one side of the fuel cell, while oxygen from the air is channeled to the cathode on the other side of the cell. At the anode, a platinum catalyst causes the hydrogen to split into positive hydrogen ions (protons) and negatively charged electrons. The PEM allows only the positively charged ions to pass through to the cathode. The negatively charged electrons must travel along an external circuit to the cathode, creating an electrical current. At the cathode, the electrons and positively charged hydrogen ions combine with oxygen to form water, which flows out of the cell.

Most fuel cells designed for use in vehicles produce less than 1.6 volts of electricity—not nearly enough to power a vehicle. Because of this, it is necessary to place multiple fuel cells onto a fuel cell stack. The potential power generated by a fuel cell stack depends on the number

A hydrogen fuel cell generates electricity through an electrochemical reaction using hydrogen and oxygen. Hydrogen is sent into one side of a proton exchange membrane (PEM). The hydrogen proton travels through the membrane, while the electron enters an electrical circuit, creating a DC electrical current. On the other side of the membrane, the proton and electron are recombined and mixed with oxygen from room air, forming pure water. Because there is no combustion in the process, there are no other emissions, making fuel cells an extremely clean and renewable source of electricity. *(Matt Stiveson, DOE/NREL)*

and size of the individual fuel cells that comprise the stack and the surface area of the PEM. Although there are not a lot of fuel cell vehicles on the road today, researchers believe they will one day revolutionize the transportation sector. The technology has the potential to significantly reduce energy use and harmful emissions as well as the U.S. dependence on foreign oil.

Fuel cell vehicles (FCVs) are an emerging technology that stands to revolutionize the transportation sector of tomorrow. A radical departure from current vehicles, they completely eliminate the conventional internal combustion engine. Like battery-electric vehicles, FCVs are propelled by electric motors. But while battery electric vehicles use electricity from an external source, FCVs create their own electricity. The fuel cell creates the electricity through a chemical process using hydrogen fuel and oxygen. FCVs can be fueled with pure hydrogen gas stored on board in high-pressure tanks. They can also be fueled with hydrogen-rich fuels such as methanol, natural gas, or even gasoline, but these fuels must first be converted into hydrogen gas by an onboard device called a reformer.

FCVs that are fueled with pure hydrogen emit no pollutants—only water and heat. FCVs using hydrogen-rich fuels and a reformer produce only small amounts of air pollutants. FCVs are twice as efficient as today's similarly sized conventional models.

Research still needs to be completed in order to begin producing FCVs for mainstream America. Effective and efficient ways to produce and store hydrogen must be determined. Currently, extensive research is under way to solve the limitations so that FCVs will become the cars of tomorrow. Partnerships with private research and the government are under way right now to make this happen with projects such as FreedomCAR (a Department of Energy (DOE) initiative) and the California Fuel Cell Partnership (a California initiative).

Zero emission hydrogen fuel cell bus at a bus stop outside a parking garage in the Connecticut transit district. This bus is nonpolluting and has more than twice the fuel economy of a standard diesel bus. *(Leslie Eudy, DOE/NREL)*

PLUG-IN VEHICLES

Plug-in hybrid technology allows gasoline-electric hybrid vehicles to be recharged from the electrical power grid and run many miles on battery power alone. Because electric motors are far more efficient than internal combustion engines, vehicles that use electricity almost always produce less global warming pollution than gasoline vehicles, even when the electricity used to fuel them is generated from coal. The benefits are even greater when vehicles are fueled with renewable generated electricity.

A gas engine provides additional driving range as needed after the battery power is gone. Plug-in hybrids may never need to run on anything but electricity for shorter commutes. The combination of gas and electric driving technologies can already achieve up to 150 MPG (241 km/gal).

In June 2008, the DOE National Renewable Energy Laboratory (NREL) modified a 2006 Toyota Prius sedan to achieve an amazing 100 MPG (161 km/gal). The experimental plug-in runs the initial 60

NREL's plug-in hybrid electric vehicle (PHEV) at the National Wind Technology Center (NWTC). This car gets 100 miles (160 km) per gallon. *(Mike Linenberger, DOR/NREL)*

miles (97 km) mostly on battery, with the remainder achieved under engine power. According to the NREL, the sedan's performance more than doubles the fuel economy of a standard Prius, which is rated 48/45 MPG. In addition, it is a fivefold improvement over the 20 MPG averages that passenger cars and light trucks in the United States achieved in 2007. The plug-in hybrid runs on electricity at low speeds; then the batteries and the gasoline engine share the work. The batteries recharge automatically as the car is running.

NREL researchers added several features to the plug-in Prius to break the 100-MPG barrier:

- A plug to recharge its batteries directly from the utility grid using a standard 110-volt electrical outlet
- A larger lithium-ion battery that allows the car to operate on electricity for longer trips at speeds up to 35 MPH
- A rooftop solar panel that charges the battery while the car is moving or parked outdoors, adding five miles to the vehicle's range

"The stored power in the battery does a great job of displacing petroleum," said Tony Markel, a senior engineer with the Vehicle Systems Analysis Group. "Most people's daily commute is about 30 miles, so this car would run virtually on battery for their entire drive."

The NREL Prius is a unique research prototype and is not available to the public. It costs about $70,000—the cost of the standard Prius plus $42,500 for the modifications.

AIR-POWERED VEHICLES

Air-powered cars are vehicles that are being developed that run on only compressed air. This zero-emission fuel is believed to hold some promise for future car models and is being explored in Europe, Asia, and the United States. Air power can be substantial as seen in pneumatic air-powered tools.

According to an article in *Green Car Journal*, a car powered by air is a reality. In fact, *compressed air vehicles*—commonly referred to as air cars—have been running around for several years. Compressed air is used every day to perform difficult tasks. For instance, mechanics rely on air-driven pneumatic tools every day to turn nuts and bolts. Pneumatic tools are powerful—even at a relatively low pounds per square inch (psi) pressure setting. Compressed air is a force that, with enough power, can propel a wheel-driven car.

The inventor Guy Negre of Motor Development International (MDI) has developed just such an air-powered vehicle. The vehicle has no combustion—its power comes solely from compressed air run by electricity from the grid. In order to make the vehicle go, a pair of air-driven pistons turns a crankshaft that produces a rotational force. The technology can potentially be paired in two-, four-, or six-cylinder engine configurations. While there is no combustion, the only engine heat comes from friction, enabling it to be constructed of lightweight aluminum.

In 2007, Tata Motors licensed the rights from MDI for $28 million to build and sell Tata air cars in India. The Nano is Tata's scooter replacement. Very popular in developing countries, it will not meet U.S. federal emissions and safety requirements. In the United States, a company called Zero Pollution Motors (ZPM) has licensed the rights to produce the MDI design in a U.S. factory. Based in New York, ZPM has an ambitious goal of rolling out a North American compressed-air vehicle for $18,000 by 2010. The company most recently unveiled MDI's newest car at the Automotive X-Prize exhibit at the New York Auto Show. ZPM and MDI will present two models—the U.S. production six-seat, four-door prototype and a three-seat, two-door economy model in the alternative class.

Air cars have not escaped the attention of U.S. automakers. Ford, for instance, has worked with an engineering team at UCLA to develop an air hybrid. Interestingly, air-powered cars are not a new idea. The concept predates the internal combustion engine. The author Jules Verne, in his book *Paris in the 21st Century* described a transportation system using compressed air. Today, scientists are turning his concept into reality.

CARS OF THE FUTURE

Cars of the future will run cleaner, faster, and more efficiently than ever before because technology is constantly being defined and improved by automakers and engine manufacturers and being watched over by the scientific community in a quest to fight the effects of global warming. As new technologies leap from the concept level to the drawing board into reality—like hybrid cars and hydrogen fuel cells—they continue to push the edges of today's car technology and redefine what is possible. There will no doubt be many new innovations as new discoveries are made—new elements, new technology, new mediums, new types of motion, which will also contribute to better fuel economy, wiser environmental management, and lower greenhouse gas emissions. As the general public becomes more aware of the pertinent issues concerning global warming, public demand can also play a significant role in what manufacturers supply in terms of efficiency.

Advanced materials—such as metals, polymers, composites, and intermetallic compounds—will play an important role in improving the efficiency of transportation engines and vehicles. Weight reduction is one of the most practical ways to increase the fuel economy of vehicles while reducing exhaust emissions. The less a car weighs, the better mileage efficiency it can achieve.

The use of lightweight, high-performance materials will contribute to the development of vehicles that provide better fuel economy but are comparable in size, comfort, and safety to today's vehicles. This way, making the change to increase energy efficiency will not affect the level of comfort previously enjoyed. The development of propulsion materials and technologies will help reduce costs while improving durability and efficiency. Regardless of what future surprises technology has in store, one thing is certain: The issue of global warming must be addressed and acted upon immediately if there is to be a future with choices.

Conclusions—the Future of Air Quality

In order to control global warming and stop the disastrous effects it will have if it is ignored any longer, action must be taken now. Every person on Earth must play an important role in contributing to the solution. One of the most significant ways people can have a direct impact is by using green energy sources in order to stop the enormous influx of carbon dioxide (CO_2) into the atmosphere. The air quality will be much healthier and global warming will be slowed. This chapter discusses the future and whether or not it is too late to take action. It then explores the benefits to health and the environment if action is taken. Next, it looks at why public awareness must exist as a firm foundation before any meaningful results will be realized. It then concludes with the importance of air quality management and what the future may hold.

IS IT TOO LATE?

The simple fact is—that it is too late to stop global warming completely. The Intergovernmental Panel on Climate Change (IPCC) has already come to the conclusion that the Earth's temperature has already begun to rise unnaturally and there is no way to reverse it. The 10 warmest years ever recorded have all occurred within the past 15 years, and during the 1900s the Earth's global average surface temperature rose about 1°F (0.6°C)—the largest increase in temperature to occur in any century during the last millennium. It is too late to completely stop this warming due to the lifetimes of the greenhouse gases already in the atmosphere. Some of them will persist for hundreds of years in the future. The most important thing to note, however, is that it is not too late to slow down the process of global warming and to reduce the amount of warming that ultimately takes place.

One of the most important things that needs to happen in order to effectively slow the process is to invest in the energy-efficient technologies that exist today, strive diligently to use clean energy sources, and get to the point where it is feasible to drive zero-emission vehicles. This would slow global warming and clear the air and allow the nation to wean itself off a long-term dependence on coal and oil.

One of the biggest steps everyone needs to take immediately is to reduce the use of fossil fuels. Governments have a significant responsibility to enact tighter and more efficient energy standards by promoting renewable sources of energy. Government policy can accomplish other things as well. For example, legislation to preserve forest ecosystems would protect forests, allowing them to survive as long-term storehouses for carbon.

Another major step forward that the federal government could take is to require cars to get much better gas mileage. Individual cities and counties also need to take the lead in reducing greenhouse gas emissions. Cities and counties can adopt many policies that promote energy efficiency and renewable energy resources. They can also enact municipal climate action plans that analyze a city's contribution to greenhouse gas emissions and set a goal for reducing municipal emissions. Practical measures often include making city buildings, streetlights, and even vehicles more energy efficient. They can use solar heating to heat buildings and swimming pools and methane from landfills to create electricity.

Every individual can help too. One of the most important ways is in the personal choice of which car to drive or whether to drive at all. It is important to remember that for every single gallon of gas burned, an average of 20 pounds (9 kg) of CO_2 enters the atmosphere. It helps tremendously when individuals choose to drive a highly fuel-efficient car, carpool, walk, take public transportation, or bicycle. When lighting indoors, using compact fluorescent lightbulbs uses only one-quarter of the energy and lasts five to 10 times longer than standard incandescent lightbulbs. Halogen lights are extremely inefficient and should be avoided.

THE FUTURE OF AIR POLLUTION AND CLIMATE CHANGE

According to a recent study by the National Oceanic and Atmospheric Administration (NOAA), short-lived gases and particle pollutants, which stay in the atmosphere for only days or weeks, have a greater impact on the Earth's climate than scientists originally thought. Even when pollutants are generated locally, they can still have global implications.

In the study, short-lived pollution included black carbon (soot), low-altitude ozone (smog), nitrates, and sulfates. Each type of pollutant has a different effect on the atmosphere. Sulfate particles, for example, have a cooling effect because they reflect sunlight. Black carbon has a warming effect because it absorbs heat.

Hiram "Chip" Levy, a senior research scientist at NOAA's Geophysical Fluid Dynamics Laboratory in Princeton, New Jersey, said, "Previous research suggests that the warming of the surface climate by increasing levels of long-lived greenhouse gases has been partially offset by increasing levels of those short-lived particles that reflect sunlight. This study found that over the 21st century the climate impacts of projected changes in human emissions of short-lived gases may in fact enhance global warming."

To illustrate how local pollutants can have global effects, Levy explained that in their climate model the projection of emissions and pollutant levels over Asia resulted in a rise in temperature and a decline in rainfall over the continental United States during the summer throughout the second half of this century.

Drew Shindell, a climate scientist at the Goddard Institute for Space Studies (GISS) in New York, remarked, "By 2050, projected changes in short-lived pollutant concentrations in two of the three studies we've made are responsible for approximately 20 percent of the simulated global-mean annual average warming. By 2100, changes in the levels of short-lived gases and particles could account for a significant portion of the predicted warming, due to a projected increase in black carbon and ozone and a decrease in sulfate."

In order to avoid these negative impacts in the future, the study concluded that reducing black carbon emissions in the domestic energy/power sector in Asia would have the biggest impact on avoiding excessive warming. They also concluded that a reduction in emissions from ground transportation in North America could have similar beneficial impacts.

Alice Gilliland, a physical scientist involved in the study, said, "To assess potential impacts of air quality management actions on future climate, current decision-making tools must be extended to consider local and global scales concurrently. There is a critical need for integrated decision making with respect to air quality and climate mitigation."

HEALTH AND ENVIRONMENTAL IMPACTS

Worldwide, increasing traffic and the associated air pollution and fuel consumption are becoming major problems for communities. The associated smog has also taken a toll. The burning of fossil fuels has contributed to the haze that disrupts the views of America's majestic national parks. Even worse, the use of fossil fuels has significant impacts on public health and the environment. If coal and other fossil fuels are used to generate electricity, the following conditions apply: Sulfur dioxide will cause acid rain; CO_2 emissions will contribute heavily to global warming; nitrogen oxides will contribute to smog; and man-made mercury will cause nerve damage.

For the health of the population and the environment, it is important that clean, renewable energy resources be used. Clean air is essential to life and good health. Electric power plants are the nation's largest industrial source of the pollutants that cause the above-listed ill effects. Air pollutants can cause serious health problems, including asthma, bronchitis, pneumonia, decreased resistance to respiratory infections, and even premature death. More Americans die from health problems caused by air pollution than from homicides. Studies by the Harvard School of Public Health and the American Cancer Society identify coal power as a cause of early death. If renewable energy resources are used, it is possible to reduce air pollution and its associated health hazards for the future.

The use of fossil fuels also damages the environment. One precious resource that is threatened is the Earth's freshwater. It takes 500 times more water to produce one kilowatt-hour of electricity from coal than from wind power. In addition, mining for coal contaminates small streams and rivers with sediment, minerals, and acid mine drainage. Acid rain from the burning of coal harms lakes and kills fish populations even in pristine areas.

The exploration, removal, and use of fossil fuels can cause enormous damage to the landscape. Depending upon the characteristics of the area, some scars left behind from the intrusion do not heal for generations. Coal mining alone creates pits, quarries, tailings, and leaves dramatically altered landscapes with little of the original biodiversity intact. Pollution control and strict land conservation are necessary if future generations are going to be able to enjoy a world with vital resources.

PUBLIC AWARENESS AS A FOUNDATION

It is critical that communities develop aggressive action plans and public education programs aimed at reducing local air emissions. The transportation sector, for example, accounts for 67 percent of U.S. oil consumption and is the predominant source of air pollution. More than half of this energy use is due to passenger vehicles (cars, SUVs, and other types of light trucks). The United States relies on the Middle East for 41 percent of its imported oil source. Oil imports from unstable foreign sources pose significant problems in terms of oil availability, oil price fluctuations, and international security.

Many opportunities exist for improving energy efficiency and reducing consumption through modifications in personal behavior, operating practices, and internal administrative policies. Public education and awareness are key to implementing those improvements.

What each individual does—or does not do—as it regards their driving behavior can make a significant difference.

Transportation-related expenses now represent 10 to 12 percent of an average family's total income, the third-largest expense in their budget. Still, most Americans continue to drive to work alone in an automobile designed to comfortably carry four or five people. With rising oil prices and shrinking of funds available for highway and traffic improvements, seeking and promoting alternative solutions to high-energy consumption and transportation is becoming a priority.

As more people become educated, they are better able to make wise choices. Through both their knowledge and example, they can lead others on the path to energy conservation and green living.

THE FUTURE OF AIR QUALITY MANAGEMENT

A critical consideration for the future is air quality. With ongoing conditions, such as overpopulation stressing natural resources, the burning of fossil fuels, the burning of the world's rain forests, and industrial processes, not only will global warming increase in intensity and play havoc with the environment, but without an effective form of air quality management in place, human populations and ecosystems will pay a high price in health and longevity.

In March 2004, a subcommittee that provides the U.S. Clean Air Act Advisory Committee (CAAAC) with independent advice drafted a National Research Council Report, making recommendations concerning future air quality management in the United States. The subcommittee consists of professional scientists from organizations such as the Health Effects Institute, Brookhaven National Laboratory, College of Medicine at the University of Cincinnati, Healthy Environments and Consumer Safety Branch of Health Canada, California State University, Clarkston University, Kennedy School of Government—Harvard University, Princeton University, Carnegie Mellon University, University of Utah, and the California Environmental Protection Agency.

Their goal is to develop scientific and technical recommendations for strengthening the nation's air quality management system, specifically focusing on human populations and natural ecosystems. Their recommendations are that for a quality plan to work it must be achieved at a steady, forward-looking evolution—not a rapid transformation—toward meeting long-term objectives. They also recommend that a workable plan will only succeed if it is implemented with a mix of administrative, regulatory, and legislative actions.

In their assessment, they identified areas where present air quality management under the Clean Air Act was sorely lacking effectiveness. One area is the inability to measure progress quantitatively to accurately confirm that goals are being met. They also determined that using a single pollutant approach is cumbersome and often bureaucratic, there is no focus on ecological effects, and there is no conclusive documentation that confirms that resources are being used to mitigate pollutants that pose the greatest risks.

The subcommittee has identified several challenges that the EPA will be facing in the near future as it strives to combat air pollution. One major hurdle will be assessing and protecting ecosystem health and ensuring justice for environmental infractions.

The subcommittee drafting the National Research Council Report on Air Quality Management for the CAAAC also identified as a major challenge the mitigation of intercontinental and cross-border transport. When particulates and other pollutants enter the atmosphere, they can travel great distances. For example, pollutants from Asia have been tracked to North America. Once pollutants enter major atmospheric wind systems, they can be carried globally. Another significant challenge will be to maintain a working, functioning air quality management system in the face of climate change. As the environment changes, the physical systems will also change, and scientists will be required to work very closely with policy makers.

Based on the fact that changes will occur, the subcommittee recommends that the scientific and technical capacity involved in the decision-making process must be much stronger to be able to handle what happens in the future. They also recommend that multiple spatial scales be taken into account. For example, both national and multistate control must be able to work together in a nested architecture. In this way, local and regional decisions are compatible with national decisions, enabling the effectiveness of decisions and actions. The subcommittee also recommends the enhanced protection of ecosystems and public welfare.

In order to strengthen scientific and technical capacity, they have identified seven key necessary tasks:

- improve emissions tracking
- enhance air pollution monitoring
- improve computer modeling and analysis
- enhance exposure assessment
- improve health and welfare assessment
- track implementation costs
- invest in research and human and technical resources

A study funded by the EPA and conducted by the National Academies National Research Council in January 2004 concluded that the Clean Air Act is working, but that a multipollutant, multistate focus is needed. It also determined there should be more of an emphasis placed on final results in order to meet future challenges. In other words, the focus should be on what works and what does not and, once determined, make the necessary adjustments.

They advised that much more must be done to improve the nation's ability to confront future air pollution. They also recommended that the EPA should use an approach that targets groups of pollutants rather than individual ones. The mobility of air pollution must be dealt with— how it travels from state to state and across international borders. In addition, improved tracking of emissions is needed to accurately assess what populations are at the highest risk of health problems from the ill effects of pollution.

The study also recommended that a method must be found to better measure the progress of pollution control strategies and that the implementation of air quality regulations be less bureaucratic, with a stronger emphasis on results, not the process. They concluded that the most important assets to protect were ecosystems and human life.

WHAT HAPPENS NEXT?

There is an overwhelming need to reduce greenhouse gas emissions to a sustainable level. Overcoming global warming and climate change is an enormous challenge. In fact, it may very well be the biggest challenge scientists, engineers, and society as a whole will face. In order to success-fully adapt, technology-based mitigations and solutions will be neces-sary. Keeping GHG emissions from exceeding target levels will require technological solutions, and for these to work they must have political backing and funding.

The Obama administration understands the environmental issues and backs the future direction the United States needs to take. The administration has vowed to break dependence on oil by promoting the next generation of cars and the alternative fuels they run on; by enhanc-ing U.S. energy supplies through responsible development of domestic renewable energy, fossil fuels, and advanced biofuels; and promoting energy efficiency by promoting investments that reduce energy bills in the transportation, electricity, industrial, building, and agricultural sectors. They also plan to take immediate action to reduce CO_2 pollu-tion, crack down on polluters, and place limits on other harmful GHG pollutants.

Much lies in store for the future with the major focus on technolog-ical advancements for detection, monitoring, and mitigation; research; wise energy choices; and public involvement.

Satellite Technology

According to an article on the Web site ScienceDaily, air quality fore-casts via space-borne satellites are one of the inventions that will be available soon. Research scientists at the National Aeronautics and Space Administration (NASA) are working on enabling a satellite con-figuration that can provide forecasts of air pollution near the ground where it has a direct impact on human health. They believe a system like this could be useful in efforts to improve air quality, assess the effective-ness of environmental regulations, and address the challenge of global warming.

In the United States, both NOAA and the EPA are currently work-ing on the application of satellite data in regional air quality forecast modeling. Presently, research is further along in Europe. According to Richard Engelen of the European Centre for Medium-Range Weather Forecasts in Reading, United Kingdom, "Regional modeling is already getting quite meaningful. Air quality forecasts are now possible up to a few days in advance in Europe where there has been a concerted effort to combine atmospheric composition data from satellite and ground stations into the existing weather forecast climate models."

A study was recently conducted by researchers from NASA's Langley Research Center in Hampton, Virginia, and Jet Propulsion Laboratory in Pasadena, California. They employed satellites to monitor air quality in Houston, Texas—a city with major air quality issues. Surprisingly, through data collection from NASA's Aura satellite, the researchers found that not all of the air pollution over Houston was caused by local sources. There was significant long-range transport from the Midwest and Ohio Valley. This was detected and confirmed by the satellite data in the computer model.

Rich Scheffe of the EPA said, "Satellites add considerable vertical and horizontal spatial detail, enabling a more scientifically sound way of understanding whether or not programs are making progress in reducing air pollution in the long term."

Satellite data still presents challenges that scientists must understand better. Researchers are trying to find ways to differentiate specific information, where they have not been successful thus far, such as how much of each gas exists at a specific altitude. The differentiation is most difficult at the Earth's surface—the region where it is most important. Satellite sensors still need to be improved to be able to accomplish this. The good news is, however, that advances in forecasting air quality are happening at a more rapid pace now due to the threat of global warming.

Automotive Research

The U.S. Department of Energy (DOE) Energy Efficiency and Renewable Energy Office currently operates the FreedomCAR and Vehicle Technologies Program. This program is developing more energy-efficient and environmentally friendly highway transportation technologies that will enable America to use less petroleum. Their long-term aim is to develop leapfrog technologies that will provide Americans with greater freedom of mobility and energy security, while lowering costs and reducing effects on the environment.

The goal of the program is to develop emission- and petroleum-free cars and light trucks. The program is conducting the research necessary to develop new technologies, such as fuel cells and advanced hybrid propulsion systems. By managing energy resources in this way, society will be able to cost-effectively move larger volumes of freight and greater numbers of passengers while emitting little or no pollution, with a dramatic reduction in dependence on imported oil.

The program's hybrid and vehicle systems research is done with industry partners. Automobile manufacturers and scientists from the program work together to design and test cutting-edge technologies.

Energy storage technologies, especially batteries, are also part of the program. Batteries are critical technologies for the development of advanced, fuel-efficient, light- and heavy-duty vehicles. They are in the process of developing durable and affordable batteries that cover many applications in a car's design, from start/stop to full-power hybrid electric, electric, and fuel cell vehicles. New batteries are being developed to be affordable, perform well, and be durable.

Advanced internal combustion engines are also being developed to be more efficient in light-, medium-, and heavy-duty vehicles. Along with efficiency, they are also being developed to meet future federal and state emissions regulations. The DOE believes this technology will lead to an overall improvement of the energy efficiency of vehicles. Advanced internal combustion engines may also serve as an important element in the transition to hydrogen fuel cells.

Research into materials technologies is another important component of the program. Advanced materials, including metals, polymers, composites, and intermetallic compounds, can play an important role in improving the efficiency of transportation engines and vehicles. Weight reduction is one of the most practical ways to improve fuel efficiency. The use of lightweight, high-performance materials will contribute to the development of vehicles that provide better fuel economy but are still comparable in size, comfort, and safety to today's vehicles.

Perhaps someday in the future, cars will be powered by energy from the Sun. In August 2005, students representing 18 universities in the United States and Canada competed in the North American Solar Challenge for victory in the world's longest solar car race. Although engineers say that solar cars will not be commercially available for many decades

This solar-powered race car was built in Philadelphia, Pennsylvania, by students at Drexel University. The car participated in the GM SunRayce USA, which was cosponsored by the U.S. Department of Energy. *(Finley R. Shapiro, DOE/NREL)*

to come—if ever—the cars that participated in the race illustrated to the American public recent advances in technology and demonstrated the promise of solar energy for other uses, such as solar cells for cars and solar energy for homes.

The challenge for the car-racing teams was to build a race car that ran on only 1,000 watts (equivalent to a hair dryer), cruised at highway speeds, and carried a driver 2,500 miles (4,023 km) from Austin, Texas, to Calgary, Canada. Each car was carefully designed and plastered with thousands of cells that collected energy from the Sun.

Future Energy Use

U.S. households produce 21 percent of the nation's global warming pollution. The good thing is that energy-conscious families can reduce their emissions by up to two-thirds. If every household in the United States made energy-efficient choices, it would eliminate 800 million tons (726 million metric tons) of global warming pollution, which adds up to more than the heat-trapping emissions from more than 100 countries.

New and emerging technologies can also reduce the production of heat-trapping gases. By choosing green power, electricity can be used that produces little or no global warming pollution. Purchasing green power goes a long way toward cutting heat-trapping emissions because clean energy sources emit little or no CO_2 pollution. While purchasing green power may be slightly higher in price, the benefits are plentiful. Some of the major benefits include the fact that it reduces smog, soot, mercury, and acid rain pollution. It also reduces financial risks. In contrast, future regulations, caps on greenhouse gases, and price fluctuations of fossil fuels could all increase the cost of energy. It will also create new jobs and generate income because green power sources usually rely on local labor, land, and resources, particularly in rural areas.

Electric customers have choices today as to the type of power they choose to purchase. As the public becomes more educated and aware of the global warming issues, they are beginning to purchase green power. Starting in 1998, roughly 30 million American utility customers began choosing their power suppliers. Customers in California and areas in New England could decide which company to buy their electricity from, which brand to buy, and what prices to pay. In other states, utilities offered green pricing, allowing customers to direct some of their electric bill toward clean, renewable energy, such as solar and wind power.

Today, about 75 million electricity customers in 42 states have the option to buy green power through their utility or an alternative power supplier, according to the National Renewable Energy Laboratory. The green revolution is gaining momentum. By harnessing wind, sunlight, plant matter, or heat from the Earth's core, it is possible to produce electricity in ways that help stop global warming pollution. The more in demand green power becomes, the less heat-trapping pollution will be generated.

Green energy comes from four sources: wind, solar, geothermal, and biomass. Wind turbines use strong wind to create pollution-free, renewable electricity. Wind power is already as cheap as fossil fuel–generated electricity in some places. The windier the location, the lower the cost is, and the more energy that can be produced. Wind energy is the most economical in places where average wind speed is at least 17 miles per hour (27 km/hr).

The Sun's energy can help produce electricity in two ways: photovoltaic (PV) systems and solar thermal systems. PV systems change sunlight directly into electricity. Solar thermal systems use the Sun's energy to heat a fluid that produces steam, which then turns a turbine and generator. Geothermal energy is generated by converting the hot water or steam from deep beneath the Earth's surface into electricity. Geothermal plants emit very little air pollution and have minimal impacts on the environment. They are also very economical. Currently, geothermal plants in the United States provide enough electricity to supply the homes of 3.5 million people. Known geothermal reserves and technology could supply the entire country with electricity for 30 years.

Biomass (crop parts or animal waste that can make energy) is an extremely versatile fuel source. It can provide electricity, heat buildings and factories, and power cars and trucks. Biomass is converted into a combustible gas, making it cleaner and more efficient. Biomass sources include agricultural residue, forest matter, and food processing by-products, as well as gas emitted from landfills. Though not as clean as wind and solar energy and a bit more expensive, its environmental benefits are a significant plus because biomass generates few or no heat-trapping gases.

As more of the population chooses to go green, less of an impact will be felt on the Earth's environment and natural resources. But it will take a dedicated effort from every person in order to slow the process. If collective efforts fail, then the world left to future generations could be a harsh, inhabitable one—very different from the one we enjoy. With each person contributing, significant achievements can take place. One person changing a lightbulb, commuting on public transportation, driving less, choosing a fuel-efficient car, using green power, purchasing ENERGY STAR® products, lowering thermostats, and recycling used materials may not seem like much. However, millions doing it makes a tremendous difference. It takes one step at a time, one person at a time. The end result will be cleaner air and a healthier world. And one thing is for certain, better or worse—we are all in this together.

APPENDIX A

Distribution of Total U.S. Greenhouse Gas Emissions by End-Use Sector, 2007					
GREENHOUSE GAS AND SOURCE	SECTOR				
	Residential	*Commercial*	*Industrial*	*Transportation*	*Total*
	Million Metric Tons Carbon Dioxide Equivalent				
Carbon Dioxide					
Energy-Related (adjusted)	1,261.3	1,097.7	1,655.2	1,902.5	**5,916.7**
Industrial Processes	—	—	105.1	—	**105.1**
Total CO$_2$	**1,261.3**	**1,097.7**	**1,760.3**	**1,902.5**	**6,021.8**
Methane					
Energy					
Coal Mining	—	—	71.1	—	**71.1**
Natural Gas Systems	—	—	176.6	—	**176.6**
Petroleum Systems	—	—	22.9	—	**22.9**
Stationary Combustion	10.4	0.1	0.6	—	**11.1**
Stationary Combustion: Electricity	0.1	0.1	0.1	—	**0.3**
Mobile Sources	—	—	—	5.1	**5.1**
Waste Management					
Landfills	—	169.0	—	—	**169.0**

Domestic Waste-water Treatment	—	17.4	—	—	**17.4**
Industrial Waste-water Treatment	—	—	9.3	—	**9.3**
Industrial Processes	—	—	**2.6**	—	**2.6**
Agricultural Sources					
Enteric Fermentation	—	—	138.5	—	138.5
Animal Waste	—	—	65.0	—	65.0
Rice Cultivation	—	—	9.7	—	9.7
Crop Residue Burning	—	—	1.4	—	1.4
Total Methane	**10.5**	**186.7**	**497.6**	**5.1**	**699.9**
Nitrous Oxide					
Agriculture					
Nitrogen Fertilization of Soils	—	—	229.6	—	229.6
Solid Waste of Animals	—	—	62.2	—	62.2
Crop Residue Burning	—	—	0.6	—	0.6
Energy Use					
Mobile Combustion	—	—	—	56.2	56.2
Stationary Combustion	0.9	0.3	4.4	—	5.7
Stationary Combustion: Electricity	3.4	3.3	2.6	—	9.3
Industrial Sources	—	—	14.0	—	14.0

(continues)

Distribution of Total U.S. Greenhouse Gas Emissions by End-Use Sector, 2007 *(continued)*					
GREENHOUSE GAS AND SOURCE	SECTOR				
	Residential	Commercial	Industrial	Transportation	Total
	Million Metric Tons Carbon Dioxide Equivalent				
Waste Management					
Human Sewage in Wastewater	—	6.0	—	—	6.0
Waste Combustion	—	—	—	—	0.0
Waste Combustion: Electricity	0.1	0.1	0.1	—	0.4
Total Nitrous Oxide:	**4.5**	**9.8**	**313.5**	**56.2**	**383.9**
Hydrofluorocarbons (HFCs)					
HFC-23	—	—	22.0	—	22.0
HFC-32	—	0.5	—	—	0.5
HFC-125	—	22.8	—	—	22.8
HFC-134a	—	—	—	72.7	72.7
HFC-143a	—	23.9	—	—	23.9
HFC-236fa	—	3.0	—	—	3.0
Total HFCs:	**0.0**	**50.2**	**22.0**	**72.7**	**144.9**
Perfluorocarbons (PFCs)					
CF_4	—	—	5.2	—	5.2
C_2F_6	—	—	4.2	—	4.2
$NF_3C_3F_8$ and C_4F_8	—	—	0.7	—	0.7
Total PFCs	**0.0**	**0.0**	**10.1**	**0.0**	**10.1**

Other HFCs, PFCs/PFPEs	—	6.1	—	—	6.1
Sulfur Hexafluoride (SF$_6$)					
SF$_6$: Utility	4.6	4.4	3.4	—	12.3
SF$_6$: Other	—	—	3.4	—	3.4
Total SF$_6$	**4.6**	**4.4**	**6.8**	**0.0**	**15.8**
Total Non-CO$_2$	**19.5**	**257.2**	**849.9**	**133.9**	**1,260.6**
Total Emissions	**1,280.8**	**1,354.7**	**2,610.4**	**2,036.4**	**7,282.4**
Source: EIA					

APPENDIX B

U.S. Emissions of Greenhouse Gases, Based on Global Warming Potential, 1990, 1995, and 2000–2007 (Million Metric Tons Carbon Dioxide Equivalent)

GAS	1990	1995	2000	2001	2002	2003	2004	2005	2006	2007
Carbon Dioxide	5,021.4	5,348.4	5,892.6	5,806.9	5,880.5	5,938.7	6,023.9	6,023.3	5,945.8	6,021.8
Methane	782.1	752.6	685.7	670.1	674.2	676.5	679.7	679.4	686.9	699.9
Nitrous Oxide	336.0	359.7	344.6	339.3	335.4	334.6	361.5	370.8	375.7	383.9
High-GWP Gases[a]	102.4	114.6	152.1	141.4	153.6	149.0	1 65.0	174.5	171.3	176.9
Total	**6,241.8**	**6,575.2**	**7,075.0**	**6,957.7**	**7,043.7**	**7,098.8**	**7,230.1**	**7,256.9**	**7,179.7**	**7,282.4**

[a]Hydrofluorocarbons (HFCs), perfluorocarbons (PFCs), and sulfur hexafluoride (SF_6).

Notes: Data in this table are revised from the data contained in the previous EIA report. Emissions of Greenhouse Gases in the United States, 2006, DOE/EIA-0573 (2006). (Washington, D.C., November 2007). Totals may not equal sum of components due to independent rounding.

Sources: Emissions: EIA estimates. Global Warming Potentials: Intergovernmental Panel on Climate Change. Climate Change 2007: The Physical Science Basis (Cambridge: Cambridge University Press, 2007), Web site: www.ipcc.ch/ipccreports/ar4-wg1.htm.

World Energy-Related Carbon Dioxide Emissions by Region, 1990–2030 (Million Metric Tons Carbon Dioxide, Percent Share of World Emissions)

REGION/COUNTRY	HISTORY[A]			PROJECTIONS[A]					AVERAGE ANNUAL PERCENT CHANGE, 2005–2030[B]
	1990	2004	2005	2010	2015	2020	2025	2030	
OECD									
OECD North America	**5,754**	**6,959**	**7,008**	**7,109**	**7,408**	**7,653**	**7,928**	**8,300**	**0.7**
%	**(27.1)**	**(25.7)**	**(25.0)**	**(22.9)**	**(21.6)**	**(20.7)**	**(20.0)**	**(19.6)**	**(9.1)**
United States	4,989	5,957	5,982	6,011	6,226	6,384	6,571	6,851	0.5
%	(23.5)	(22.0)	(21.3)	(19.3)	(18.1)	(17.2)	(16.6)	(16.2)	(6.1)
Canada	465	623	628	669	698	727	756	784	0.9
%	(2.2)	(2.3)	(2.2)	(2.2)	(2.0)	(2.0)	(1.9)	(1.9)	(1.1)
Mexico	300	379	398	430	484	542	601	655	2.1
%	(1.4)	(1.4)	(1.4)	(1.4)	(1.4)	(1.5)	(1.5)	(1.6)	(1.9)
OECD Europe	**4,101**	**4,373**	**4,383**	**4,512**	**4,678**	**4,760**	**4,800**	**4,834**	**0.4**
%	**(19.3)**	**(16.2)**	**(15.6)**	**(14.5)**	**(13.6)**	**(12.9)**	**(12.1)**	**(11.4)**	**(3.2)**
OECD Asia	**1,541**	**2,148**	**2,174**	**2,208**	**2,287**	**2,322**	**2,357**	**2,403**	**0.4**
%	**(7.3)**	**(7.9)**	**(7.8)**	**(7.1)**	**(6.7)**	**(6.3)**	**(6.0)**	**(5.7)**	**(1.6)**
Japan	1,009	1,242	1,230	1,196	1,201	1,195	1,184	1,170	-0.2
%	(4.8)	(4.6)	(4.4)	(3.8)	(3.5)	(3.2)	(3.0)	(2.8)	(-0.4)

(continues)

World Energy-Related Carbon Dioxide Emissions by Region, 1990–2030 (continued)

REGION/COUNTRY	HISTORY[A]			PROJECTIONS[A]					AVERAGE ANNUAL PERCENT CHANGE, 2005–2030[B]
	1990	2004	2005	2010	2015	2020	2025	2030	
OECD									
South Korea	241	488	500	559	612	632	656	693	1.3
%	(1.1)	(1.8)	(1.8)	(1.8)	(1.8)	(1.7)	(1.7)	(1.6)	(1.4)
Australia/New Zealand	291	418	444	454	474	495	517	540	0.8
%	(1.4)	(1.5)	(1.6)	(1.5)	(1.4)	(1.3)	(1.3)	(1.3)	(0.7)
Total OECD	**11,396**	**13,480**	**13,565**	**13,829**	**14,373**	**14,736**	**15,085**	**15,538**	**0.5**
%	**(53.7)**	**(49.8)**	**(48.4)**	**(44.5)**	**(41.9)**	**(39.8)**	**(38.1)**	**(36.7)**	**(13.8)**
Non-OECD									
Non-OECD Europe & Eurasia	**4,198**	**2,797**	**2,865**	**3,066**	**3,330**	**3,508**	**3,625**	**3,811**	**1.1**
%	**(19.8)**	**(10.3)**	**(10.2)**	**(9.9)**	**(9.7)**	**(9.5)**	**(9.2)**	**(9.0)**	**(6.6)**
Russia	2,376	1,669	1,696	1,789	1,902	1,984	2,020	2,117	0.9
%	(11.2)	(6.2)	(6.0)	(5.8)	(5.5)	(5.4)	(5.1)	(5.0)	(2.9)
Other	1,822	1,128	1,169	1,278	1,428	1,524	1,606	1,694	1.5
%	(8.6)	(4.2)	(4.2)	(4.1)	(4.2)	(4.1)	(4.1)	(4.0)	(3.7)
Non-OECD Asia	**3,613**	**7,517**	**8,177**	**10,185**	**12,157**	**13,907**	**15,683**	**17,482**	**3.1**
%	**(17.0)**	**(27.8)**	**(29.2)**	**(32.7)**	**(35.4)**	**(37.6)**	**(39.6)**	**(41.3)**	**(65.2)**
China	2,241	4,753	5,323	6,898	8,214	9,475	10,747	12,007	3.3
%	(10.6)	(17.6)	(19.0)	(22.2)	(23.9)	(25.6)	(27.1)	(28.4)	(46.8)

India	565	1,127	1,164	1,349	1,604	1,818	2,019	2,238	2.6
%	(2.7)	(4.2)	(4.1)	(4.3)	(4.7)	(4.9)	(5.1)	(5.3)	(7.5)
Other Non-OECD Asia	807	1,637	1,690	1,938	2,338	2,614	2,917	3,237	2.6
%	(3.8)	(6.0)	(6.0)	(6.2)	(6.8)	(7.1)	(7.4)	(7.6)	(10.8)
Middle East	**700**	**1,290**	**1,400**	**1,622**	**1,802**	**1,988**	**2,120**	**2,250**	**1.9**
%	**(3.3)**	**(4.8)**	**(5.0)**	**(5.2)**	**(5.4)**	**(5.4)**	**(5.4)**	**(5.3)**	**(6.0)**
Africa	**649**	**943**	**966**	**1,090**	**1,244**	**1,366**	**1,450**	**1,515**	**1.8**
%	**(3.1)**	**(3.5)**	**(3.4)**	**(3.5)**	**(3.6)**	**(3.7)**	**(3.7)**	**(3.6)**	**(3.8)**
Central and South America	**669**	**1,042**	**1,078**	**1,308**	**1,429**	**1,531**	**1,628**	**1,729**	**1.9**
%	**(3.2)**	**(3.8)**	**(3.8)**	**(4.2)**	**(4.2)**	**(4.1)**	**(4.1)**	**(4.1)**	**(4.6)**
Brazil	216	350	356	451	498	541	582	633	2.3
%	(1.0)	(1.3)	(1.3)	(1.5)	(1.5)	(1.5)	(1.5)	(1.5)	(1.9)
Other Central/ South America	453	692	722	857	931	990	1,046	1,097	1.7
%	(2.1)	(2.6)	(2.6)	(2.8)	(2.7)	(2.7)	(2.6)	(2.6)	(2.6)
Total Non-OECD	**9,830**	**13,589**	**14,486**	**17,271**	**19,962**	**22,299**	**24,506**	**26,787**	**2.5**
%	**(46.3)**	**(50.2)**	**(51.6)**	**(55.5)**	**(58.1)**	**(60.2)**	**(61.9)**	**(63.3)**	**(86.2)**
TOTAL WORLD	**21,266**	**27,070**	**28,051**	**31,100**	**34,335**	**37,035**	**39,591**	**42,325**	**1.7**

[a]Values adjusted for nonfuel sequestration.

[b]Values in parentheses indicate percent share of total world absolute change.

Note: The U.S. numbers include carbon dioxide emissions attributable to renewable energy sources.

Sources: History: Energy Information Administration (EIA), International Energy Annual 2005 (May–July 2007), Web site: www.eia.doe. gov/iea/ and data presented in this report. Projections: EIA, Annual Energy Outlook 2008, DOE/EIA-0383 (2008) (Washington, D.C., June 2008), Table 1, Web site www.eia.doe.gov/oiaf/aeo; and International Energy Outlook 2008, DOE/EIA-0484(2008) (Washington, D.C. September 2008). Table A10.

Source: EIA

# CHRONOLOGY

ca. 1400–1850   Little Ice Age covers the Earth with record cold, large glaciers, and snow. There is widespread disease, starvation, and death.

1800–70   The levels of CO_2 in the atmosphere are 290 ppm.

1824   Jean-Baptiste Joseph Fourier, a French mathematician and physicist, calculates that the Earth would be much colder without its protective atmosphere.

1827   Jean-Baptiste Joseph Fourier presents his theory about the Earth's warming. At this time many believe warming is a positive thing.

1859   John Tyndall, an Irish physicist, discovers that some gases exist in the atmosphere that block infrared radiation. He presents the concept that changes in the concentration of atmospheric gases could cause the climate to change.

1894   Beginning of the industrial pollution of the environment.

1913–14   Svante Arrhenius discovers the greenhouse effect and predicts that the Earth's atmosphere will continue to warm. He predicts that the atmosphere will not reach dangerous levels for thousands of years, so his theory is not received with any urgency.

1920–25   Texas and the Persian Gulf bring productive oil wells into operation, which begins the world's dependency on a relatively inexpensive form of energy.

1934   The worst dust storm of the dust bowl occurs in the United States on what historians would later call Black Sunday. Dust storms are a product of drought and soil erosion.

1945   The U.S. Office of Naval Research begins supporting many fields of science, including those that deal with climate change issues.

1949–50 Guy S. Callendar, a British steam engineer and inventor, propounds the theory that the greenhouse effect is linked to human actions and will cause problems. No one takes him too seriously, but scientists do begin to develop new ways to measure climate.

1950–70 Technological developments enable increased awareness about global warming and the enhanced greenhouse effect. Studies confirm a steadily rising CO_2 level. The public begins to notice and becomes concerned with air pollution issues.

1958 U.S. scientist Charles David Keeling of the Scripps Institution of Oceanography detects a yearly rise in atmospheric CO_2. He begins collecting continuous CO_2 readings at an observatory on Mauna Loa, Hawaii. The results became known as the famous Keeling Curve.

1963 Studies show that water vapor plays a significant part in making the climate sensitive to changes in CO_2 levels.

1968 Studies reveal the potential collapse of the Antarctic ice sheet, which would raise sea levels to dangerous heights, causing damage to places worldwide.

1972 Studies with ice cores reveal large climate shifts in the past.

1974 Significant drought and other unusual weather phenomenon over the past two years cause increased concern about climate change not only among scientists but with the public as a whole.

1976 Deforestation and other impacts on the ecosystem start to receive attention as major issues in the future of the world's climate.

1977 The scientific community begins focusing on global warming as a serious threat needing to be addressed within the next century.

1979 The World Climate Research Programme is launched to coordinate international research on global warming and climate change.

1982   Greenland ice cores show significant temperature oscillations over the past century.

1983   The greenhouse effect and related issues get pushed into the political arena through reports from the U.S. National Academy of Sciences and the Environmental Protection Agency.

1984–90   The media begins to make global warming and its enhanced greenhouse effect a common topic among Americans. Many critics emerge.

1987   An ice core from Antarctica analyzed by French and Russian scientists reveals an extremely close correlation between CO_2 and temperature going back more than 100,000 years.

1988   The United Nations set up a scientific authority to review the evidence on global warming. It is called the Inter-governmental Panel on Climate Change (IPCC) and consists of 2,500 scientists from countries around the world.

1989   The first IPCC report says that levels of human-made greenhouse gases are steadily increasing in the atmosphere and predicts that they will cause global warming.

1990   An appeal signed by 49 Nobel prizewinners and 700 members of the National Academy of Sciences states, "There is broad agreement within the scientific community that amplification of the Earth's natural greenhouse effect by the buildup of various gases introduced by human activity has the potential to produce dramatic changes in climate . . . Only by taking action now can we insure that future generations will not be put at risk."

1992   The United Nations Conference on Environment and Development (UNCED), known informally as the Earth Summit, begins on June 3 in Rio de Janeiro, Brazil. It results in the United Nations Framework Convention on Climate Change, Agenda 21, the Rio Declaration on Environment and Development Statement of Forest Principles, and the United Nations Convention on Biological Diversity.

1993 Greenland ice cores suggest that significant climate change can occur within one decade.

1995 The second IPCC report is issued and concludes there is a human-caused component to the greenhouse effect warming. The consensus is that serious warming is likely in the coming century. Reports on the breaking up of Antarctic ice sheets and other signs of warming in the polar regions are now beginning to catch the public's attention.

1997 The third conference of the parties to the Framework Convention on Climate Change is held in Kyoto, Japan. Adopted on December 11, a document called the Kyoto Protocol commits its signatories to reduce emissions of greenhouse gases.

2000 Climatologists label the 1990s the hottest decade on record.

2001 The IPPC's third report states that the evidence for anthropogenic global warming is incontrovertible, but that its effects on climate are still difficult to pin down. President Bush declares scientific uncertainty too great to justify Kyoto Protocol's targets.

The United States Global Change Research Program releases the findings of the National Assessment of the Potential Consequences of Climate Variability and Change. The assessment finds that temperatures in the United States will rise by 5 to 9°F (3–5°C) over the next century and predicts increases in both very wet (flooding) and very dry (drought) conditions. Many ecosystems are vulnerable to climate change. Water supply for human consumption and irrigation is at risk due to increased probability of drought, reduced snow pack, and increased risk of flooding. Sea-level rise and storm surges will most likely damage coastal infrastructure.

2002 Second hottest year on record.

Heavy rains cause disastrous flooding in Central Europe leading to more than 100 deaths and more than $30 billion in damage. Extreme drought in many parts of the world (Africa, India,

Australia, and the United States) results in thousands of deaths and significant crop damage. President Bush calls for 10 more years of research on climate change to clear up remaining uncertainties and proposes only voluntary measures to mitigate climate change until 2012.

2003   U.S. senators John McCain and Joseph Lieberman introduce a bipartisan bill to reduce emissions of greenhouse gases nation-wide via a greenhouse gas emission cap and trade program.

Scientific observations raise concern that the collapse of ice sheets in Antarctica and Greenland can raise sea levels faster than previously thought.

A deadly summer heat wave in Europe convinces many in Europe of the urgency of controlling global warming but does not equally capture the attention of those living in the United States.

International Energy Agency (IEA) identifies China as the world's second largest carbon emitter because of their increased use of fossil fuels.

The level of CO_2 in the atmosphere reaches 382 ppm.

2004   Books and movies feature global warming.

2005   Kyoto Protocol takes effect on February 16. In addition, global warming is a topic at the G8 summit in Gleneagles, Scotland, where country leaders in attendance recognize climate change as a serious, long-term challenge.

Hurricane Katrina forces the U.S. public to face the issue of global warming.

2006   Former U.S. vice president Al Gore's *An Inconvenient Truth* draws attention to global warming in the United States.

Sir Nicholas Stern, former World Bank economist, reports that global warming will cost up to 20 percent of worldwide gross domestic product if nothing is done about it now.

2007   IPCC's fourth assessment report says glacial shrinkage, ice loss, and permafrost retreat are all signs that climate change is underway now. They predict a higher risk of drought, floods, and more powerful storms during the next 100 years. As a result, hunger, homelessness, and disease will increase. The atmosphere may warm 1.8 to 4.0°C and sea levels may rise 7 to 23 inches (18 to 59 cm) by the year 2100.

Al Gore and the IPCC share the Nobel Peace Prize for their efforts to bring the critical issues of global warming to the world's attention.

2008　The price of oil reached and surpassed $100 per barrel, leaving some countries paying more than $10 per gallon.

ENERGY STAR® appliance sales have nearly doubled. ENERGY STAR® is a U.S. government-backed program helping businesses and individuals protect the environment through superior energy efficiency.

U.S. wind energy capacity reaches 10,000 megawatts, which is enough to power 2.5 million homes.

2009　President Obama takes office and vows to address the issue of global warming and climate change by allowing individual states to move forward in controlling greenhouse gas emissions. As a result, American automakers can prepare for the future and build cars of tomorrow and reduce the country's dependence on foreign oil. Perhaps these measures will help restore national security and the health of the planet, and the U.S. government will no longer ignore the scientific facts.

The year 2009 will be a crucial year in the effort to address climate change. The meeting on December 7–18 in Copenhagen, Denmark, of the UN Climate Change Conference promises to shape an effective response to climate change. The snapping of an ice bridge in April 2009 linking the Wilkins Ice Shelf (the size of Jamaica) to Antarctic islands could cause the ice shelf to break away, the latest indication that there is no time to lose in addressing global warming.

# GLOSSARY

aerosols   tiny bits of liquid or solid matter suspended in air. They come from natural sources such as erupting volcanoes and from waste gases emitted from automobiles, factories, and power plants. By reflecting sunlight, aerosols cool the climate and offset some of the warming caused by greenhouse gases.

alternative fuels   also known as nonconventional fuels, any materials or substances that can be used as a fuel, other than conventional fuels. Conventional fuels include: fossil fuels (petroleum (oil), coal, propane, and natural gas), and nuclear materials such as uranium. Some well-known alternative fuels include biodiesel, bioalcohol (methanol, ethanol, butanol), chemically stored electricity (batteries and fuel cells), hydrogen, nonfossil methane, nonfossil natural gas, vegetable oil, and other biomass sources.

anthropogenic   made by people or resulting from human activities. This term is usually used in the context of emissions that are produced as a result of human activities.

atmosphere   the thin layer of gases that surrounds the Earth and allows living organisms to breathe. It reaches 400 miles (644 km) above the surface, but 80 percent is concentrated in the troposphere—the lower seven miles (11 km) above the Earth's surface.

biodiversity   different plant and animal species.

biofuels   solid, liquid, or gaseous fuels derived from relatively recently dead biological material, distinguished from fossil fuels, which are derived from long-dead biological material. Theoretically, biofuels can be produced from any (biological) carbon source, although, the most common sources are photosynthetic plants. Various plants and plant-derived materials are used for biofuel manufacturing.

biomass   plant material that can be used for fuel.

cap-and-trade   the cap-and-trade system involves the trading of emission allowances, where the total allowance is strictly limited or capped. Emissions trading is an administrative approach used to control pollution by providing economic incentives for achieving reductions in the emissions of pollutants. A company is allowed to have a specified level of pollution; they can sell and trade. If a company exceeds their limit, they can buy credits to decrease global warming elsewhere.

carbon   a naturally abundant nonmetallic element that occurs in many inorganic and in all organic compounds.

carbon capture and storage   a method of mitigating the contribution of fossil fuel emissions to global warming, based on capturing CO_2 from large point sources such as coal-fired power plants. The goal is to store the CO_2 permanently away from the atmosphere, typically in underground geologic formations.

carbon cycle   a colorless, odorless gas that forms when carbon atoms combine with oxygen atoms. Carbon dioxide is a tiny, but vital, part of the atmosphere. The heat-absorbing ability of carbon dioxide is what makes life possible on Earth.

carbon dioxide   a colorless, odorless gas that passes out of the lungs during respiration. It is the primary greenhouse gas and causes the greatest amount of global warming.

carbon sequestration   a geoengineering technique for the long-term storage of CO_2 for the mitigation of global warming. It is usually captured through chemical, biological, or physical processes.

carbon sink   an area where large quantities of carbon are built up in the wood of trees, in calcium carbonate rocks, in animal species, in the ocean, or any other place where carbon is stored. These places act as reservoirs, keeping carbon out of the atmosphere.

Clean Air Act   One of a number of pieces of legislation relating to the reduction of smog and air pollution in general. The use by governments to enforce clean air standards has contributed to an improvement in human health and longer life spans.

climate   the usual pattern of weather that is averaged over a long period of time.

climate model   a quantitative way of representing the interactions of the atmosphere, oceans, land surface, and ice. Models can range from relatively simple to extremely complicated.

climatologist   a scientist who studies the climate.

compressed-air vehicle   a new experimental car that uses compressed air, as opposed to the gas-and-oxygen explosions of internal-combustion models.

concentration   the amount of a component in a given area or volume. In global warming, it is a measurement of how much of a particular gas is in the atmosphere compared to all of the gases in the atmosphere.

condense   the process that changes a gas into a liquid.

daylighting   The effective use of natural light in a building in order to reduce the need for artificial (electrical) lighting, which also improves the visual quality of an area and saves energy resources.

deforestation   the large-scale cutting of trees from a forested area, often leaving large areas bare and susceptible to erosion.

desertification   the process that turns an area into a desert.

ecological   the protection of the air, water, and other natural resources from pollution or its effects. It is the practice of good environmentalism.

ecosystem   a community of interacting organisms and their physical environment.

electric vehicle   An electric vehicle (EV) is a vehicle with one or more electric motors for propulsion.

emissions   the release of a substance (usually a gas when referring to the subject of climate change) into the atmosphere.

energy efficiency   using less energy to provide the same level of energy service. An example would be insulating a home to use less heating and cooling energy to achieve the same temperature.

evaporation   the process by which a liquid, such as water, is changed to a gas.

feedback a change caused by a process that, in turn, may influence that process. Some changes caused by global warming may hasten the process of warming (positive feedback); some may slow warming (negative feedback).

flexible fuel vehicle an alternative fuel vehicle with an internal combustion engine designed to run on more than one fuel, usually gasoline blended with either ethanol or methanol fuel, and both fuels are stored in the same common tank.

food chain the transfer of food energy from producer (plant) to consumer (animal) to decomposer (insect, fungus, etc.).

forcing mechanisms that disrupt the global energy balance between incoming energy from the Sun and outgoing heat from the Earth. By altering the global energy balance, such mechanisms force the climate to change. Today, anthropogenic greenhouse gases added to the atmosphere are forcing climate to warm.

fossil fuels an energy source made from coal, oil, or natural gas. The burning of fossil fuels is one of the chief causes of global warming.

fuel cells an electrochemical conversion device. It produces electricity from fuel (on the anode side) and an oxidant (on the cathode side), which react in the presence of an electrolyte. The reactants flow into the cell, and the reaction products flow out of it, while the electrolyte remains within it. Fuel cells can operate virtually continuously as long as the necessary flows are maintained.

fuel efficiency the efficiency of a process that converts chemical potential energy contained in a carrier fuel into kinetic energy or work.

glacier a mass of ice formed by the buildup of snow over hundreds and thousands of years.

global warming an increase in the temperature of the Earth's atmosphere, caused by the buildup of greenhouse gases. This is also referred to as the enhanced greenhouse effect caused by humans.

global warming potential (GWP) a measure of how much a given mass of greenhouse gas is estimated to contribute to global warming. It is a relative scale, which compares the gas in question to that of the same mass of carbon dioxide, whose GWP is equal to 1.

green building　a green building, or sustainable building, is an outcome of a design that focuses on increasing the efficiency of resource use—energy, water, and materials—while reducing building impacts on human health and the environment during the building's life cycle, through better siting, design, construction, operation, maintenance, and removal.

greenhouse effect　the natural trapping of heat energy by gases present in the atmosphere, such as carbon dioxide, methane, and water vapor. The trapped heat is then emitted as heat back to Earth.

greenhouse gas　a gas that traps heat in the atmosphere and keeps the Earth warm enough to allow life to exist.

hybrid vehicle　a vehicle that uses two or more distinct power sources to move the vehicle. The term most commonly refers to hybrid electric vehicles (HEVs), which combine an internal combustion engine and one or more electric motors.

industrial revolution　the period during which industry developed rapidly as a result of advances in technology. This started in Britain during the late 18th and early 19th centuries.

IPCC (Intergovernmental Panel on Climate Change)　This UN organization consists of 2,500 scientists that assesses information in the scientific and technical literature related to the issue of climate change. The United Nations Environment Programme and the World Meteorological Organization established the IPCC jointly in 1988.

land use　the management practice of a certain land cover type. Land use may be such things as forest, arable land, grassland, urban land, and wilderness.

land use change　an alteration of the management practice on a certain land cover type. Land use changes may influence climate systems because they affect evapotranspiration and sources and sinks of greenhouse gases. An example of land use change is removing a forest to build a city.

methane　a colorless, odorless, flammable gas that is the major ingredient of natural gas. Methane is produced wherever decay occurs and little or no oxygen is present.

nitrogen as a gas, nitrogen takes up 80 percent of the volume of the Earth's atmosphere. It is also an element in substances such as fertilizer.

nitrous oxide a heat-absorbing gas in the Earth's atmosphere. Nitrous oxide is emitted from nitrogen-based fertilizers.

organic Rankine cycle a Rankine cycle is a closed-circuit steam cycle. An organic Rankine cycle uses a heated chemical instead of steam. Chemicals used in the organic Rankine cycle include Freon, butane, propane, ammonia, and the new environmentally friendly refrigerants.

ozone a molecule that consists of three oxygen atoms. Ozone is present in small amounts in the Earth's atmosphere at 14 to 19 miles (23–31km) above the Earth's surface. A layer of ozone makes life possible by shielding the Earth's surface from most harmful ultraviolet rays. In the lower atmosphere, ozone emitted from auto exhausts and factories is an air pollutant.

ozone precursor gases gases that react to form ozone.

parts per million (ppm) the number of parts of a chemical found in 1 million parts of a particular gas, liquid, or solid.

permafrost permanently frozen ground in the Arctic. As global warming increases, this ground is melting.

photosynthesis the process by which plants make food, using light energy, carbon dioxide, and water.

protocol the terms of a treaty that have been agreed to and signed by all parties.

radiation the particles or waves of energy.

renewable something that can be replaced or regrown, such as trees, or a source of energy that never runs out, such as solar energy, wind energy, or geothermal energy.

resources the raw materials from the Earth that are used by humans to make useful things.

satellite any small object that orbits a larger one. Artificial satellites carry instruments for scientific study and communication. Imagery taken from satellites is used to monitor aspects of global warming, such as glacier retreat, ice cap melting, desertification, erosion, hurricane damage, and flooding. Sea-surface temperatures and measurements are also obtained from man-made satellites in orbit around the Earth.

simulation   a computer model of a process that is based on actual facts. The model attempts to mimic, or replicate, actual physical processes.

thermal   something that relates to heat.

tropical   a region that is hot and often wet (humid). These areas are located around the Earth's equator.

troposphere   the bottom layer of the atmosphere, rising from sea level up to an average of about 7.5 miles (12 km).

tundra   a vast treeless plain in the Arctic with a marshy surface covering a permafrost layer.

weather   the conditions of the atmosphere at a particular time and place. Weather includes such measurements as temperature, precipitation, air pressure, and wind speed and direction.

FURTHER RESOURCES

BOOKS

Christianson, Gale. *Greenhouse: The 200-Year Story of Global Warming.* New York: Walker, 1999. This book looks at the enhanced greenhouse effect worldwide after the industrial revolution and outlines the consequences to the environment.

Dow, Kirstin, and Thomas E. Downing. *The Atlas of Climate Change: Mapping the World's Greatest Challenge.* Los Angeles: University of California Press, 2006. This publication offers maps and geographic statistics and information on climate change, global warming, economics, and other related scientific topics worldwide.

Friedman, Katherine. *What If the Polar Ice Caps Melted?* Danbury, Conn.: Children's Press, 2002. This book focuses on environmental problems related to the Earth's atmosphere, including global warming, changing weather patterns, and their effects on ecosystems.

Gelbspan, Ross. *The Heat Is On: The High Stakes Battle over Earth's Threatened Climate.* Reading, Mass.: Addison Wesley, 1997. This work offers a look at the controversy environmentalists often face when they deal with fossil fuel companies.

Harrison, Patrick, Gail McLeod, and Patrick G. Harrison. *Who Says Kids Can't Fight Global Warming.* Chattanooga, Tenn.: Pat's Top Products, 2007. This book offers real solutions that everybody can do to help solve the world's biggest air pollution problems.

Houghton, John. *Global Warming: The Complete Briefing.* New York: Cambridge University Press, 2004. This book outlines the scientific basis of global warming and describes the impacts that climate change will have on society. It also looks at solutions to the problem.

Langholz, Jeffrey. *You Can Prevent Global Warming (and Save Money!): 51 Easy Ways.* Riverside, N.J.: Andrews McMeel Publishing, 2003. This book aims at converting public concern over global

warming into positive action to stop it by providing simple, every-
day practices that can easily be done to minimize it, as well as save
money.

McKibben, Bill. *Fight Global Warming Now: The Handbook for Taking
Action in Your Community.* New York: Holt Paperbacks, 2007. This
book provides the facts of what must change to save the climate. It
also shows how everyone can act proactively in their community to
make a difference.

Pringle, Laurence. *Global Warming: The Threat of Earth's Changing Climate.* New York: SeaStar Publishing Company, 2001. This
book provides information on the carbon cycle, rising sea levels, El
Niño, aerosols, smog, flooding, and other issues related to global
warming.

Ruddiman, William F. *Earth's Climate: Past and Future.* New York:
W. H. Freeman and Company, 2001. This book takes a detailed look
at the history of the Earth's climate and the forces that have shaped
it over time.

Thornhill, Jan. *This Is My Planet—the Kids Guide to Global Warming.*
Toronto, Ontario: Maple Tree Press, 2007. This book offers students
the tools they need to become ecologically oriented by taking a
comprehensive look at climate change in polar, ocean, and land-
based ecosystems.

Weart, Spencer R. *The Discovery of Global Warming (New Histories of
Science, Technology, and Medicine).* Cambridge, Mass.:Harvard University Press, 2004. This book traces the history of the global warming concept through a long process of incremental research rather
than a dramatic revelation.

JOURNAL ARTICLES

Allen, Jeannie. "Tango in the Atmosphere: Ozone and Climate
Change." *NASA Research Features, NASA Earth Observatory* (February 2004). Available online. URL: www.giss.nasa.gov/research/
features/tango/?print=1&1=1&2=2&3=3. Accessed January 25,
2009. This article discusses the interdependence of ozone and

global warming and what the results will be if the release of the greenhouse gases that cause destruction of the ozone layer are not stopped.

Andrews, Wyatt. "Clean Coal—Pipe Dream or Next Big Thing?" CBSNews.com. (6/20/08). Available online. URL: www.cbsnews. com/stories/2008/06/20/evingnews/main4199506.shtml. Accessed February 22, 2009. This articles discusses the process used for clean coal technology and whether it is a legitimate answer to global warming.

Barboza, David, and Keith Bradsher. "Pollution from Chinese Coal Casts a Global Shadow." *New York Times* (6/11/06). Available online. URL: www.nytimes.com/2006/06/11/business/worldbusiness/llchi-nacoal.html. Accessed January 13, 2009. This article discusses the global effects of air pollution caused by the burning of coal.

Borenstein, Seth. "Carbon Dioxide Emissions up 3 Percent in '07." Ajc.com. (10/26/08). Available online. URL: www.ajc.com/green/content/printedition/2008/09/26/emissions.html. Accessed January 1, 2009. This article presents evidence supporting the conclusion that CO_2 levels are continuing to climb despite the Kyoto.

Bradsher, Keith. "Push to Fix Ozone Layer and Slow Global Warming." *New York Times* (3/15/07). Available online. URL: http://www.nytimes.com/2007/03/15/business/worldbusiness/15warming.html. Accessed January 23, 2009. This article looks at an international meeting held in Hong Kong in March 2007 to put stricter limits on the use of refrigerants for air-conditioners.

———. "China's Boom Adds to Global Warming Problem." Health and Energy. (10/22/03). Available online. URL: http://healthandenergy.com/China_burning_more_coal.htm. Accessed January 16, 2009. This article discusses the threat to global warming that China will play now and in the immediate future using coal.

Bradsher, Keith and David Barboza. "Pollution from Chinese Coal Casts a Global Shadow." *New York Times* (6/11/06). Available online. URL: www.nytimes.com/2006/06/11/business/Worldbusiness/11chinacoal/html. Accessed June 7, 2009. This article discusses

China's rapid industrialization and the worldwide effects of its pollution.

Broder, John M. "Obama Affirms Climate Change Goals." *New York Times* (11/19/08). Available online. URL: http://www.nytimes.com/2008/11/19/us/politics/19climate.html?_r=1&pagewanted=print. Accessed January 27, 2009. This article outlines Obama's plans to deal with global warming and the role the United States must play in the future in order to solve the problem.

Carey, Bjorn. "Soot Could Hasten Melting of Arctic Ice." Live Science (3/28/05). Available online. URL: www.livescience.com/environment/050328_arctic_soot.html. Accessed January 26, 2009. This article talks about computer modeling using NASA data to determine how the ice caps will react to pollution in the face of global warming.

Choi, Charles Q. "The Energy Debates: Clean Coal." LiveScience (12/5/08). Available online. URL: www.ivescience.com/environment/081205-energy-debates-clean-coal.html. Accessed February 22, 2009. This article discusses whether or not clean coal technology up to its expectations.

Economist. "Coal Power, Still Powerful." (11/15/07). Available online. URL: www.economist.com/business/displaystory.cfm?story_id=10145492. Accessed January 10, 2009. This article discusses the still-strong dependence on coal worldwide.

Environment Maine. "Coal Rush to Threaten Environment—Over 150 Proposed Plants Would Boost Global Warming Pollution by 10 Percent, Coal Consumption by 30 Percent; Dirty Technologies Predominate." (7/20/06). Available online. URL: www.environmentmaine.org/news-releases/global-warming/global-warming. Accessed. January 22, 2009. This discusses various U.S. companies still interested in relying heavily on old coal energy generation technology.

Fackler, Martin. "Latest Honda Runs on Hydrogen, Not Petroleum." *New York Times* (6/17/08). Available online. URL: www.nytimes.com/2008/06/17/business/17fuelcell.html?fta=y. Accessed January 29, 2009. This article discusses Honda's 16-year research program to develop a hydrogen-powered vehicle that can now be mass-produced.

Gorman, Steve. "California Pursues Low-Carbon Fuel Constraints." (4/13/09). Available online. URL: www.planetark.com/enviro-news/item/52562. Accessed June 12, 2009: This article discusses the low-carbon fuel standards that California is expected to adopt in order to curb carbon emissions from transportation fuels.

Greenhouse, Linda. "Justices Say E.P.A. Has Power to Act on Harmful Gases." *New York Times* (4/3/07). Available online. URL: www.nytimes.com/2007/04/03/washington/03scotus.html?pagewanted=print. Accessed January 21, 2009. This article details how the U.S. Supreme Court has ruled that the EPA has the authority to regulate greenhouse gas emissions in automobile emissions.

Kahn, Joseph, and Mark Landler. "China Grabs West's Smoke-Spewing Factories." *New York Times* (12/21/07). Available online. URL: www.nytimes.com/2007/12/21/world/asia/21transfer.html?pagewanted=print. Accessed January 26, 2009. This article discusses pollution problems in China's key manufacturing industries and their onward industrial expansion.

Kaho, Todd. "Trends: Air Powered Cars." *Green Car Journal* (5/27/08). Available online. URL: www.greencar.com/articles/trends-air-powered-cars.php. Accessed February 28, 2009. This article discusses the evolution of futuristic air-powered cars that use no fuels.

Llanos, Miguel. "Jump Start for Solar? Car Race Shows Potential." MSNBC (8/5/05). Available online. URL: http://www.msnbc,msn.com/id/8737930. Accessed March 16, 2009. This article highlights solar-powered race cars in the North American Solar Challenge.

Lohr, Steve. "Energy Standards Needed, Report Says." *New York Times* (5/17/07). Available online. URL: www.nytimes.com/2007/05/17/business/17energy.html?_r=1&pagewanted=print. Accessed January 26, 2009. This article discusses the new product standard mandates that will be needed for current technology's energy saving opportunities.

Madrigal, Alexis. "China's 2030 CO_2 Emissions Could Equal the Entire World's Today." Wired Science (2/8/08). Available online. URL: http://blog.wired.com/wiredscience/2008/02/chinas-2030.co2.html. Accessed January 15, 2009. This article explores the explosive industrialization occurring in China and their heavy use of coal-generated energy.

Mapes, Jennifer. "UN Scientists Warn of Catastrophic Climate Changes." National Geographic News (2/06/01). Available online. URL: http://news.nationalgeographic.com/news/2001/02/0206_climate1.html. Accessed January 15, 2009. This article discusses the IPCC's findings on global warming.

Mayell, Hillary. "Soot Identified as Major Contributor to Global Warming." National Geographic News (2/12/01). Available online. URL: http://news.nationalgeographic.com/news/2001/02/0212_climate2.html. Accessed January 15, 2009. This article looks at pollutants as a factor in global warming.

Moore, Frances C. "Carbon Dioxide Emissions Accelerating Rapidly." Earth Policy Institute (4/9/08). Available online. URL: www.earth-policy.org/Indicators/CO2/2008.htm. Accessed January 21, 2009. This article discusses why CO_2 emissions are rising rapidly and why fossil fuels are to blame.

O' Driscoll, Patrick, and Dan Vergano. "Fossil Fuels Are to Blame, World Scientists Conclude." *USA Today* (3/1/07). Available online. URL: www.usatoday.com/tech/science/2007-01-30-ipcc-report_x.htm. Accessed January 13, 2009. This article examines the role that fossil fuels have played in the global warming situation.

Revkin, Andrew D. "U.S. Predicting Steady Increase for Emissions." *New York Times* (3/3/07). Available online. URL: www.nytimes.com/2007/03/03/science/03climate.html. Accessed January 26, 2009. This article discusses the projections the U.S. government is making as to what they expect future greenhouse gas contributions will be.

———. "Stuck on Coal, and Stuck for Words in a High-Tech World." *New York Times* (12/4/07). Available online. URL: www.nytimes.com/2007/12/04/science/earth/04comm.html?pagewanted=print. Accessed January 15, 2009. This article discusses the persistent use of coal as an energy choice.

———. "Obama: Climate Plan Firm Amid Economic Woes." *New York Times* (11/18/08). Available online. URL: http://dotearth.blogs. nytimes.com/2008/11/18/obama-climate-message-amid-economic- woes/. Accessed January 26, 2009. This article outlines Obama's viewpoint on global warming and his plans to curb heat-trapping gases and lessen U.S. dependence on foreign oil.

———. "Industry Ignored Its Scientists on Climate." *New York Times* (4/24/09). Available online. URL: www.nytimes.com/2009/04/24/ science/earth/24deny.html. Accessed June 10, 2009. This article discusses how scientists of the Global Climate Coalition—a group working for fossil fuel industry and trying to discredit the notion of anthropogenically caused global warming—deliberately suppressed from public disclosure the written opinion of their very own research scientists that there was indeed unequivocal evidence of an anthropogenic element to the issue. Instead, their chosen stance was that there was still too much uncertainty to justify strict regulation of CO_2 emissions.

Rosenthal, Elisabeth. "China Increases Lead as Biggest Carbon Dioxide Emitter." *New York Times* (06/14/08). Available online. URL: www.nytimes.com/2008/06/14/world/asia/14china. html?_r=1&pagewantted=print. Accessed January 21, 2009. This article addresses China's own industrial revolution and the global consequences.

———. "Europe Turns Back to Coal, Raising Climate Fears." *New York Times* (4/23/08). Available online. URL: www.nytimes. com/2008/04/23/world/europe/23coal.html?_r=1&ref=science& pagewanted=print. Accessed January 26, 2009. Discusses Italy's current plans to increase their reliance on coal over the next five years.

ScienceDaily. "Stratospheric Injections to Counter Global Warming Could Damage Ozone Layer." Available online. URL: www.science- daily.com/releases/2008/04/080424140407.htm. Accessed April 25, 2008. This article discusses the possibility of using geoengineering to counteract the global warming process.

———. "Air Quality Forecasts See Future in Space." (12/19/07). Available online. URL: http://www.sciencedaily.com/releases/2007/12/07

1213101348.htm. Accessed March 5, 2009. This article discusses the use of satellite technology to monitor and forecast air pollution.

———. "Greenhouse Gases, Carbon Dioxide, and Methane Rise Sharply in 2007." (4/24/08). Available online. URL: www.science-daily.com/releases/2008/04/080423181652.htm. Accessed January 2, 2009. This article discusses the continual rise in carbon dioxide levels in the atmosphere since the industrial revolution.

Stark, Anne M. "Modeling of Long-term Fossil Fuel Consumption Shows 14.5-degree Hike in Earth's Temperature." Lawrence Livermore National Laboratory (11/1/05). Available online. URL: http://public affairs.llnl.gov/news_releasees/2005/NR-05-11-01p.html. Accessed January 13, 2009. This article looks at the possible outcome of world climate if reductions in fossil fuel use are not undertaken now.

Stern, Nicholas. "Time for a Green Industrial Revolution." *New Scientist.* magazine. Available online. URL: www.newscientist.com/article/mg20126926.600-comment-time-for-a-green-industrial-revolution. Accessed February 1, 2009. This discuses the economics and feasibility of phenomena on scientific theory.

Terhune, Lea. "Worldwide Impact from Climate Change Predicted." USINFO (4/06/07). Available online. URL: http://usinfo.state.gov/utils/printpage.html. Accessed January 21, 2009. This article discusses global impacts such as drought, desertification, glacial retreat, communicable diseases, and impacts to natural systems and ecosystems as a result of unchecked global warming.

Thompson, Andrea. "Two Evils Compete: Global Warming vs. Ozone Hole." LiveScience (4/24/08). Available online. URL: www.livescience.com/environment/080424-sulfur-ozone-hole.html. Accessed January 23, 2009. This article discusses the interaction between the ozone hole and global warming and discusses a suggested technique of injecting sulfur into the atmosphere to cool the temperatures and slow global warming.

Unger, Nadine. "Interaction of Ozone and Sulfate in Air Pollution and Climate Change." NASA Science Briefs, Goddard Institute for Space Studies (March 2006). Available online. URL: www.giss.nasa.gov/research/features/tango/?print=1&1=1&2=2&3=3. Accessed March

13, 2008. This article discusses the complementary effects that ozone and climate have on each other and what it means to global warming and future populations.

UPI. "Energy Resources: China to Build New Coal Plants." (12/17/08). Available online. URL: www.upi.com/Energy_Resources/2008/12/17/China_to_build_new_coal_plants/UPI. Accessed January 16, 2009. This article addresses the heavy dependence China has on coal-powered energy and how continued growth will contribute heavily to global warming.

Venkataraman, Bina. "The Other Global Warming." *Boston Globe* (1/25/09). Available online. URL: www.boston.com/bostonglobe/ideas/articles/2009/01/25/the_other_global_warming/. Accessed February 20, 2009. This articles discusses other types of environmental warming that may become as important in the near future as global warming is today.

Wald, Matthew L. "Cleaner Coal Is Attracting Some Doubts." *New York Times* (2/21/07). Available online. URL: www.nytimes.com/2007/02/21/business/21coal.html?_r=1&pagewanted=print. Accessed January 26, 2009. This article discusses why some environmentalists are worried about new gasification technology for coal plants being able to live up to their expectations.

Walli, Ron. "CO_2 Emissions Booming, Shifting East, Researchers Report." Oak Ridge National Laboratory Public Release (9/24/08). Available online. URL: www.eurekalert.org/pub_releases/2008-09/drnl-ceb092408.php. Accessed January 2, 2009. This article discusses the steady increase in CO_2 emissions from China.

WEB SITES
Global Warming

BBC: Climate Change. Sponsored by the BBC Weather Centre, UK. Available online. URL: www.bbc.co.uk/climate/. Accessed December 31, 2008. Provides evidence, impacts, adaptation, policies, and links about climate change.

Climate Ark home page. Sponsored by Ecological Internet. Available online. URL: www.climateark.org. Accessed October 23,

2007. Promotes public policy that addresses global climate change through reduction in carbon and other emissions, energy conservation, alternative energy sources, and ending deforestation.

Climate Solutions home page. Sponsored by Atmosphere Alliance and Energy Outreach Center. Available online. URL: www.climatesolutions.org. Accessed October 23, 2007. Offers practical solutions to global warming.

Environmental Defense Fund home page. Sponsored by Environmental Defense Fund. Available online. URL: www.environmentaldefense.org. Accessed October 26, 2007. From an organization started by a handful of environmental scientists in 1967 that provides quality information and helpful resources on understanding global warming and other crucial environmental issues.

Environmental Protection Agency home page. Sponsored by the U.S. Environmental Protection Agency. Available online. URL: www.epa.gov. Accessed October 26, 2007. Provides information about EPA's efforts and programs to protect the environment. It offers a wide array of information on global warming.

European Environment Agency home page. Sponsored by the European Environment Agency in Copenhagen, Denmark. Available online. URL: www.eea.europa.eu/themes/climate. Accessed October 26, 2007. Posts their reports on topics such as air quality, ozone depletion, and climate change.

Global Warming FAQs. Sponsored by the National Oceanic and Atmospheric Administration. Available online. URL: http://lwf.ncdc.noaa.gov/oa/climate/globalwarming.html. Accessed December 31, 2008. Presents basic questions and answers about global warming.

Global Warming: Focus on the Future home page. Sponsored by EnviroLink. Available online. URL: www.enviroweb.org. Accessed October 26, 2007. Offers statistics and photography of global warming topics.

HotEarth.Net home page. Sponsored by National Environmental Trust. Available online. URL: www.net.org/warming. Accessed October 26, 2007. Features informational articles on the causes of global warming, its harmful effects, and solutions that could stop it.

Intergovernmental Panel on Climate Change (IPCC) home page. Sponsored by the World Meteorological Organization (WMO) and the United Nations Environment Programme (UNEP). Available online. URL: http://www.ipcc.ch/. Accessed October 26, 2007. Offers current information on the science of global warming and recommendations on practical solutions and policy management.

NASA's Goddard Institute for Space Studies home page. Sponsored by the National Aeronautics and Space Administration. Available online. URL: www.giss.nasa.gov. Accessed October 26, 2007. Provides a large database of information, research, and other resources.

NOAA's National Climatic Data Center home page. Sponsored by the National Oceanic and Atmospheric Administration. Available online. URL: www.ncdc.noaa.gov. Accessed October 26, 2007. Offers a multitude of resources and information on climate, climate change, global warming.

Ozone Action home page. Sponsored by the Southeast Michigan Council of Governments. Available online. URL: www.semcog. org/services/ozoneaction/kids.htm. Accessed October 26, 2007. Provides information on air quality by focusing on ozone, the atmosphere, environmental issues, and related health issues.

Perlman, David. "New Global Warming Evidence Presented." *San Francisco Chronicle* (2/19/05). Available online. URL: www.sfgate. com/cgi-bin/article.cgi?file=/c/a/2005/02/19/MNGE1BECPI1.DTL. Accessed February 2, 2009. Discusses the conclusive evidence that the Scripps Institution of Oceanography has collected that proves without a doubt global warming exists.

Scientific American home page. Sponsored by Scientific American, Inc. Available online. URL: www.sciam.com. Accessed October 23, 2007. Offers an online magazine and often presents articles concerning climate change and global warming.

Tyndall Centre at University of East Anglia home page. Sponsored by the Tyndall Centre for Climate Change Research. Available online. URL: http://www.tyndall.ac.uk. Accessed October 26, 2007. Offers information on climate change and is considered one of the leaders in UK research on global warming.

UK's Climatic Research Unit. Sponsored by the Climate Research Unit, School of Environmental Sciences, University of East Anglia, Norwich, UK. Available online. URL: www.cru.uea.ac.uk/. Aims to "improve scientific understanding of past climate history and its impact on humanity, the course and causes of climate change during the present century, and future prospects."

Union of Concerned Scientists home page. Sponsored by the Union of Concerned Scientists. Available online. URL: www.ucsusa.org. Accessed October 26, 2007. Offers quality resource sections on global warming and ozone depletion.

United Nations Environment Programme–World Meteorological Organization. Sponsored by the United Nations Environment Programme. Available online. URL: www.gcrio.org/ipcc/qa/index.htm. Accessed December 31, 2008. Contains an easy-to-follow question and answer presentation on global warming.

United Nations Framework Convention on Climate Change (UNFCCC) home page. Sponsored by the United Nations Framework Convention on Climate Change. Available online. URL: http://unfccc.int/2860.php. Accessed October 26, 2007. Presents a spectrum on climate change information and policy.

U.S. Global Change Research Program home page. Sponsored by the U.S. Office of Science and Technology Policy, the Office of Management and Budget, and the Council on Environmental Quality. Available online. URL: www.usgcrp.gov. Accessed October 26, 2007. Provides information on the current research activities of national and international science programs that focus on global monitoring of climate and ecosystem issues.

World Wildlife Foundation Climate Change Campaign home page. Sponsored by the World Wildlife Fund. Available online. URL: www.worldwildlife.org/climate/. Accessed October 26, 2007. Contains information on what various countries are doing, and not doing, to deal with global warming.

Greenhouse Gas Emissions

Energy Information Administration home page. Sponsored by the U.S. Department of Energy. Available online. URL: www.eia.doe. gov/environment.html. Accessed October 26, 2007. Lists official environmental energy-related emissions data and environmental analyses from the U.S. government. This site contains U.S. carbon dioxide, methane, and nitrous oxide emissions data and other greenhouse reports.

World Resources Institute—Climate, Energy & Transport home page. Sponsored by the World Resources Institute. Available online. URL: www.wri.org/climate/publications.cfm. Accessed October 26, 2007. Offers a collection of reports on global technology deployment to stabilize emissions, agriculture, and greenhouse gas mitigation, climate science discoveries, and renewable energy.

INDEX

Italic page numbers indicate illustrations or maps. Page numbers followed by *c* denote chronology entries; page numbers followed by *t* denote tables, charts, or graphs.

A

acid rain 49, 51, 68, 72–77, *75*, 81*c*, 214
active systems 94
AEI (automotive environmental index) 121, 122
aerosols 13
Africa 15–16, 193
AFVs (alternative fueled vehicles) 118–122
Agricultural Research Service, USDA (ARS) 187
agriculture 27–28, 149–157, *153, 154, 156*
Agriculture, U.S. Department of 180
AgSTAR program 150
air-conditioning 61
air inversions 55–56, *57, 58,* 70
air pollution
  acid rain 72–77, *75*
  agriculture 149–151
  cars/trucks 108–109, 115–118
  China 45, 160
  coal plants 77–79
  Donora smog 68, 69
  future issues 212–213
  and global dimming 15–19
  and health xv, 213–214
  London smog 68–72, *71*
  oil 12
  power plants 137
  satellite modeling 218–219
  since industrial revolution 62–86
  soot and particles *16,* 138–140, 143
Air Pollution Control Act (1955) 73
air-powered vehicles 207–208
air quality management 215–217
Allen, Russell 90–91
alternative fueled vehicles (AFVs) 118–122
alternative transportation fuels 174–186
  biodiesel 155, 176–177, 176*t,* 191

electricity 177
ethanol. *See* ethanol
hydrogen *120,* 182–183
methanol 130, 180–181, *181,* 190
natural gas 183–185
propane 185–186
American Petroleum Institute 143, 146
anaerobic decomposition/digestion 27, 125, 127, 149, 157, 188
anaerobic digester 150
Andersen, Jennifer 45–47
anthracite coal 3
anthropogenic greenhouse effect 6, 34, *35,* 57*t*
Arctic 139–140, 195
Arrhenius, Svante 79, 81*c,* 232*c*
ARS (USDA Agricultural Research Service) 187
Asia 16, 59, 60, 140, 212, 213
asthma 54–55
atmospheric brown cloud xvi, 14, 45, 59, 60, 140, 161
Aurilio, Anna 169
automobile manufacturers 162, 163
automobiles. *See* cars
automotive environmental index (AEI) 121, 122

B

bacteria 11, 28
bagasse 154
Bala, Govindasamy 20, 21
Balbus, John 54–55
Balmes, John 54
Barnett, Tim 137
batteries 220
battery-electric vehicle (BEV) 177, 201–202, *202*
bicycles 115, *115,* 132
biobutanol 186–187
biodiesel 155, 176–177, 176*t,* 191
biodiversity 75
biofuels 152, 172. *See also specific fuels, e.g.:* ethanol
biogas 150, 157, 188–189
biomass